THE
internet
PLAYGROUND

Toby Miller
General Editor

Vol. 10

PETER LANG
New York • Washington, D.C./Baltimore • Bern
Frankfurt am Main • Berlin • Brussels • Vienna • Oxford

ELLEN SEITER

THE
internet
PLAYGROUND

Children's Access,
Entertainment,
and Mis-Education

PETER LANG
New York • Washington, D.C./Baltimore • Bern
Frankfurt am Main • Berlin • Brussels • Vienna • Oxford

Library of Congress Cataloging-in-Publication Data

Seiter, Ellen.
The Internet playground: children's access, entertainment,
and mis-education / Ellen Seiter.
p . cm. — (Popular culture & everyday life; v. 10)
Includes bibliographical references and index.
1. Internet and children. I. Title. II. Series.
HQ784.I58S46   004.67'8'083—dc22   2004027257
ISBN 0-8204-7124-0
ISSN 1529-2428

Bibliographic information published by **Die Deutsche Bibliothek**.
Die Deutsche Bibliothek lists this publication in the "Deutsche
Nationalbibliografie"; detailed bibliographic data is available
on the Internet at http://dnb.ddb.de/.

Cover design by Lisa Barfield

The paper in this book meets the guidelines for permanence and durability
of the Committee on Production Guidelines for Book Longevity
of the Council of Library Resources.

© 2005 Peter Lang Publishing, Inc., New York
275 Seventh Avenue, 28th Floor, New York, NY 10001
www.peterlangusa.com

Printed in the United States of America

For Anne and Henry and Joe
Together we're unlimited

WITHDRAWN

Contents

Illustrations

Acknowledgments

Scott Kessler's record in community organizing is a testament to how much can be accomplished through imaginative thinking, a tireless commitment to activism, and an interest in promoting the social good over making profits. Scott sparked the idea of the computer class, developed the new school facility, and did the formidable political work that brought this project to fruition. As CEO of the San Diego Business Improvement District Council, he and his staff secured a broad coalition of support for this project, including the local small business community, the City Council, and the local schools. The after-school program depended on the constant support of my administrator Judy Moore. Nathan Price kept the project afloat with his grant writing. Marco Anguiano secured the newspaper printing, layout, and distribution we needed. I am also indebted to the guest speakers in the Washington class: Sara-Ellen Amster, Blues for the Schools, Giovanna Chesler, Zeinabu Irene Davis, Susan G. Davis, DeeDee Halleck, Dan Hallin, George Lipsitz, Jane Rhodes, Michael Schudson, and the staff of UCSD Community Pediatrics. Thanks are also due to: Kevin Bentz, Shawna Caballero, Cynthia Chris, Judy Elliott, Julia Himberg, Bruce Jones, Marie Judson, Matt Ratto, Jennifer Sahm, Catherine Saulino, Liz Sisco, and Jennifer Wenn.

Initial funding for this project was provided by the University of California Office of the President Urban-School Collaborative research grant. It is my hope that this book stands as evidence of the continuing promise and urgent need for more community K–12 programs, at a time when outreach budgets have been slashed and University of California education falls further out of reach for many working-class students.

Additional research assistance came from the UCSD Committee on Research and the UCSD Civic Collaborative. Vivian Reznik of the Community Pediatrics program at the UCSD Medical School provided crucial support through her grant with the California Department of Health.

All of the computers for the class were purchased with a generous grant from the Price-Weingart Fund of the San Diego Foundation. Additional support came from the Vons Foundation and the Ralphs Food-4-Less Foundation, the San Diego Commission on Arts and Culture, the San Diego Business Improvement District Council, and the Adams Avenue Business Association.

My brother Charles Seiter kept me alert to the important questions regarding the Internet and computers. I was profoundly influenced by conversations with Susan Davis, Olga Vasquez, Anita Schiller, and Michael Cole, and by the writing of Ann Haas Dyson, David Buckingham, and Clifford Stoll. George Lipsitz encouraged me at every stage of the project. Dan Schiller understood in the mid-1990s the role that traditional entertainment entities would play in the development of the Internet—his conversation, research help, and moral support were tremendously valuable. Toby Miller provided encouragement at just the right moment. Damon Zucca played a key role in seeing the book to completion.

Thanks to my sister, Rosemary Morrison, for being my first and best teacher. Anne, Henry, and Joe Metcalf challenged my thinking day by day and made the research process lively and fun. Anne came to the rescue when my energies were flagging with her skillful editing. Henry took a keen interest in my students and helped me decode unfamiliar aspects of popular culture. From kindergarten through third grade, Joe good-naturedly accepted babysitters so that I could teach the after school class. My own children's heartfelt support for my involvement in the computer class means more to me than I can say.

chapter 1

Children's Use of Computers at Home and at School

Long ago—back in the Internet's ancient history, like 1994 or 1995—some Internet enthusiasts were so captivated by the technology's powers they regarded "cyberspace" as the end of for-profit communication. . . . That was then.

—ROBERT McCHESNEY (1997, 33)

Technological revolutions are not the same as social revolutions and are more likely, in our times, to be the opposite. But they do have this in common: they do not simply happen but must be made to happen.

—DAVID NOBLE (1984, 195)

The technology gap between rich and poor American children is widening. The growing importance of the Internet has created a new disparity across class lines in children's access to skills, social networks, and intellectual resources. While children with high household incomes enjoy speedy, pervasive access to technology at home and at school (either private or affluent public school districts), others struggle to compete for intermittent access to slow machines that are outdated and erratic. While corporations scheme to profit from children's interest in Internet use, they target children in different ways.

Despite these disparities in quality and quantity of access, children of all social classes rapidly find engaging materials on the Internet and master the technical skills necessary to use the Web for game playing and downloading of pictures and audio and video files. They gravitate toward play on the World Wide Web, and find much to their satisfaction: games, music, fan materials. Corporations seek to capitalize on children's fascination and proclivity for Internet use, but they follow different accumulation strategies with regard to the children's market. One large group, the mass market, is targeted as consumers of publicity and potential audiences in the expanding global entertainment market, or as purchasers of small-ticket branded goods such as soft drinks, candy, and chips. This type of marketing strives to reach those with disposable income regardless of whether they live in India, the Czech Republic, or Brazil. A different kind of marketing targets affluent families and school districts with expensive high-end products, such as online courses, educational software, and subscription-based Web content. Mass marketers of junk food and entertainment properties strive to entice poorer children with free videos and music, with sweepstakes and contests, and with fan promotions.

From 1999 to 2003, I volunteered to teach children at two public elementary schools in southern California. One, which I will call Clearview, is at the high end of technology and Internet access; the other, which I will call Washington, is a struggling, technology-poor school. I taught weekly after-school classes in their (very different) computer labs, and assisted small groups of children on a pull-out basis for computer exercises. My observations of children in these two school environments form the basis for this book and for my analysis of the disparities among children with regard to Internet access. The two schools in my study are located in the same southern California city—a location with a large high technology sector—and serve the same primary school grades. They are separated geographically by twenty miles and socially by a chasm of difference with regard to school funding, teacher qualification, and class size, as well as parents' income, education, and levels of everyday stress.

Educational Marketing and Electronic Exploitation

Clearview serves grades K–6 and about seven hundred students. In each and every classroom from kindergarten on, a bank of three or four computers are available for children to use at the teacher's discretion. In addition, there are twelve to fifteen computers in the school library—or "media center," as school libraries are now called—where the book collection is online, and various encyclopedic programs and grade-specific educational themes are available from the district server. In addition to classroom and library computers, there is a dedicated computer classroom with a full-time teacher or two, where every class is brought as a group to complete a project once a week. The computer curriculum is carefully planned to accommodate the students' grade level and to ensure that they can complete a task and emerge with a printed example of their project at the end of the class time. For younger grades, this might mean creating digital designs of snowflakes in winter or pumpkins at Halloween; for older grades it might involve making crossword puzzles or history timelines based on their social studies projects. Printers flow freely at each and every location, so that the children bring home to parents each week material proof of their developing computer literacy. Keyboarding is emphasized early on, with the explicit goal of making each child proficient in touch typing by the fifth grade.

An expert group of parent fundraisers (many of them mothers who formerly worked as business professionals but now devote the bulk of their time to home-making and child rearing) works throughout the year to continually equip the school with technology (on the treadmill of planned obsolescence) and to pay for extra teachers the district cannot afford. The school's well-administered foundation designated technology the top funding priority from 1995 through 2003; as the state's budget crisis worsened, additional art, science, and music teachers moved higher on the list.

The school technology teacher runs an open lab time during lunch for especially eager students. Most of the students enjoy the use of up-to-date computer technology at home, too. Typical families in the neighborhood have three or four computers at home, and it is not uncommon for children in first grade to have their own computers in their private bedrooms. In this respect, they outpace their European counterparts in the construction of "media-rich" bedrooms (Livingstone 2002). Digital cameras, graphics software, and Web design are common hobbies among the boys. By fourth grade, these specially experienced "tech gods" often demonstrate better troubleshooting skills than their teachers (Cuban 2001). Local students are sufficiently enamored of the new computers installed yearly that they were rumored to have twice stolen groups of new flat-screen computers from the media center.

At each of the school's computer sites, the desks are arranged in clover patterns, making eye contact impossible unless students reposition their chairs and move away from their own screens. Indeed, I was very surprised at how rarely students assisted one another at computer tasks. Yet the design of the personal computer and the spatial arrangement of these labs make peer interaction difficult. Wandering off task to search other areas of the computer is prevented by the program, and children may not exit the worksheet or the program before their task is completed. For the most part, Clearview students sit silently at their stations, headphones on, locked out of interaction with anything except the screen. Once completed, the student's performance is recorded centrally, enumerating his or her score, comparing it to last week's score and calculating the time taken to answer each question. While each student is presumably pitted only against themselves and their own log of work, the children frequently peek around to their neighbors' screens to compare speed and progress, to see how fast they are coming along or which worksheet they are working on. At Clearview, adults supervise closely to keep children on task, and security software on the machines prevents them from frivolous Web surfing.

Along with the children's enjoyment of this technology-rich environment comes early experience with software systems (adopted from employee management tools) that regulate, monitor, and control young students. As kindergartners, children are brought to the school's media center to use an educational program served up from the school district's central facilities. The system is called "SuccessMaker" and is produced by the Computer Curriculum Corporation. The school's technology coordinator assigns each child a username and password, so that each time they log on to the school's network their work, the time it took to complete it, and their computer activities are monitored. Under the watchful eye of their teacher, a parent volunteer, the librarian, or the technology specialist, students usually complete a single exercise in math or reading skills, designated for their ability level, and are timed in its execution. After each question in reading or math, they can find out whether they were right or wrong and how many incorrect answers they've given so far. With each correct answer they are rewarded with a little animation of a cute animal. If they are slow to reply, the cartoon creature appears on the screen to ask kindly if they need help. Wrong answers are greeted with cheery messages such as "Try again!" The educational software in use restricts children's activities to exercises pitched precisely to the school's curricular standards and to the individual's precise achievement to date, based on grade-level norms. The company promises individualized instruction: one of their slogans is "Changing the world. One learner at a time." But in fact the software enforces rigid adherence to only the content that appears on the curricular standards.

Computer Curriculum Corporation, or CCC, and its product SuccessMaker have been very successful during a decade when hundreds of such products failed on the market. Introduced in 1993, SuccessMaker promised to accurately predict the precise amount of time it would take for a student to meet curricular standards. CCC was founded in 1967 by a group of Stanford University professors—including Richard Atkinson, who eventually became chancellor of the University of California system from 1998 to 2003. CCC belongs to an era before venture capital flooded the software market, when the costs of computers and networks were still prohibitive for elementary schools and homes (Harris 1993). Launched with federal money from the Office of Education, which was interested in exploring computerized instruction, CCC expanded with interest-free loan money from Hewlett Packard, which was also interested in exploring the educational market.

At first CCC's growth was slow, but by 1990 schools' investment in computers was on the rise, and CCC was bought for $75 million by Simon and Schuster, for Paramount Communications. The educational market, its expansion boosted both by privatization and by technology acquisitions, was now seen as a surefire winner. For a firm such as Paramount, with 40 percent of its revenues coming from publishing, and 47 percent of those from education, educational software represented a chance to repurpose content in a seemingly endless number of versions. "We'll sell the content twice and three times and four times," boasted Richard Snyder, CEO of Simon and Schuster, specifying that educational content was seen as especially "amenable to electronic exploitation" (Weber 1993). CCC sells Internet-based software to school districts at a cost of thousands per site license. By 2002, SuccessMaker was offering six thousand hours of electronic instructions for grades pre-K through eight, and was used by more than 4.5 million students (NCS Learn Home Page 2002).

Critics expressed doubts in the 1980s that the kinds of programs offered by CCC were anything more than electronic worksheets, and claimed that they rewarded rote memorization rather than critical thinking (Berliner & Biddle 1995; McNeil 2000; Kohn 2000). I was shocked by the unimaginative and deeply traditional methods of learning SuccessMaker promoted. The program seemed to have achieved a nearly exact replication of the boring and conservative content, artwork, and questions found in the routine worksheets distributed for decades to U.S. students. But there is one undeniable advantage to SuccessMaker over the books of reproducible paper worksheets sold to teachers at a fraction of the cost: SuccessMaker delivers concrete, quantifiable, individualized results. These results can be aggregated by teachers (for evaluation purposes) by grade level, by school (for budgeting and personnel decisions), and by school district—where standardized measures of achievement can be mobilized to justify all sorts of administrative decisions. In southern California, as in upscale markets across the United States, these kinds of aggregated data on student performance relative to grade level are widely publicized so that prospective residents can gauge the precise level of achievement of each neighborhood—and high-performing schools can drive a wildly inflated real estate market.

While many of district teachers privately told me that they were less than thrilled with SuccessMaker, especially for young children, parents and administrators found it highly attractive. It is surprising how favorably parents at Clearview responded to the SuccessMaker program: parents often request that their children be allowed to spend extra time after or before school working with the program, convinced that it is the key to improved grades. They volunteer to supervise open lab time so that their own children can have extra access. In the intensely competitive atmosphere of the school, other parents then complained to school officials about the perceived unfair advantage additional time on SuccessMaker might give students. These parental complaints (as well as the efforts of parents to bring children in before the start of the school day to use SuccessMaker) testify to their belief that mere contact with drill-and-practice software would magically improve children's performance.

As computer scientist Clifford Stoll sardonically describes this ideal of cyber-school: "Yep, just sign up for the future: the parent-pleasin', tax-savin', teacher-firin', interactive-educatin', child-centerin' Cyberschool. No stuffy classrooms. No more teacher strikes. No outdated textbooks. No expensive clarinet lessons. No boring homework. No learning. Coming soon to a school district near you" (1999, 107). The incursion of drill-and-practice software into schools goes hand in glove with the curricular emphasis on standardized testing as the sole mark of educational quality and excellence. In fact, computers coach students in the form of standardized testing: fill in the blank, multiple choice, and true/false. Education researchers refer to this type of computer education as "drill and kill" for its negative impact on student enthusiasm for learning.

Instead of promoting challenging, complex forms of knowledge and thinking in schools, computers have, in the opinion of many educators, encouraged simplistic models of learning. Most forms of computer education do little to advance critical thinking skills, but instead consist of electronic worksheets that offer practice in the format of the standardized, computer-scorable test itself. David Buckingham suggests that "while children understand the potential of computers in principle, they rarely engage in relatively more creative or technically complex activities themselves tending as a result to under-use the potential of the computer quite considerably" (2002, 7). The educational benefits of computing—divorced from socioeconomic advantages—are unknown. Although it is widely assumed to be true, no research proves that familiarity with computers and the Internet increases learning and competitiveness in the job market. Even among students most highly skilled in using digital media, weariness is growing. Graduate degree programs are filling with students who were early, avid fans of the Web, studied programming in college, and gained employment in the Silicon Valley and its many rival technology regions, only to find the work unsatisfying and seek a second career. The jobs that the high-tech sector provides are both fewer in number and far less desirable than many imagined in the 1990s: "Internet employment is cyclical, insecure, and often unrewarding. Today's digital sweatshop is a cubicle, with tight deadlines and often no promise of work beyond the end of each month . . . by holding the computer as provider of both education and employment, we're grossly overselling technology and causing an absurd overspecialization in our workforce" (Stoll 1999, 174). Why then would a school like Clearview adopt expensive software like SuccessMaker? At the level of school administration, "enterprise computing" such as that promoted by CCC, promises to streamline the data-gathering needs of all compartments of the school. Rather than individualized record keeping of grades, the performance of students can be continually monitored and linked to attendance records and family information.

> The most important application of enterprise computing is improving student achievement. The standards movement, in particular, has spurred the development of enterprise technologies to help educators to this end. Standards-based education holds out the expectation that teachers should be able to conduct continuous, formative assessment and offer students appropriate instruction that matches their needs and abilities. Enterprise computing, with its emphasis on data management, can offer these capabilities. Additionally, enterprise technology can help aggregate classroom level data so administrators can know which classrooms or schools need additional resources in order to meet local and state standards. (Milone 1999, 28)

This account of enterprise computing emphasizes the positive spin. Like so many modern management discourses, it hides the substantially punitive use of all these calculations. Teachers, pupils, and entire neighborhoods are labeled "low-performing" by these tools, wrongly imposing a business efficiency model on the notoriously inefficient process of educating children. Teachers are denied pay increases, schools lose funding, and students' confidence is completely undermined by the type of calculations these software programs provide.

Some of the harshest critics of public education in recent years have been executives of prominent high technology corporations. Kenneth Saltman explains the cynicism of such advocates of school reform:

> Obviously, the "quality" issue that corporate CEOs such as Louis Gerstner of IBM hurl at the public schools do not concern the schools from which IBM will draw employees. Those schools—heavily funded, largely white schools in suburbs—are providing IBM with well-educated members of the professional class. So the "failing" schools are not the ones from which IBM intends to benefit by hiring their graduates. However, IBM can certainly benefit from convincing the government that IBM can allow urban schools and their students to compete with wealthy white suburban schools for IBM's shrinking number of jobs if those urban schools buy IBM's many products. (2002, 253)

CCC is now controlled by Pearson, publisher of London's *Financial Times*, and owner of the world's largest online business diploma operation, and "the world's largest education company." Theirs is the big business model of education: Pearson claims to sell one in four textbooks in use, and to be the leading English-language learning publisher in the world, supplying 120 million textbooks and programs around the world each year.

Pearson contends that the global education market reached a worth of a billion dollars per year in 2003 (Pearson 2002). The head of Pearson Education, Peter Jovanovich, remained optimistic in 2004, "as the recovery in state budgets and federal No Child Left Behind funds help our testing and digital learning businesses" (Milliot 2004).

Reverse Heritage

Two things stand out as remarkable from teaching children about computers and the Internet skills. Children are fearless in their interaction with the technology, and they swiftly acquire new computer skills, even when they are encountering them for the first time. Adults born before the personal computer boom of the 1980s stand in awe of children today and find our children's superior comfort with and mastery of computers to be stupefying. Adults, especially women and those with less education, may avoid learning about computers due to a fear of technology, a fear of being made to look stupid before others (especially other family members), or a concern that they will damage the machine or erase all of the data. An ethnographic study of those without a home computer in San Diego found that nonusers frequently thought of computers as meant for yuppies, college students, geeks, or businesspeople (Stanley 2001, 7). Latinos—the largest group in the community served by my computer lab at Washington—appear most reluctant of all ethnic groups. While we experimented with a variety of efforts—from open houses to summer hours—to draw adult learners into the computer lab, each one met with failure. Many Latina mothers made sacrifices to ensure their children's attendance at the lab, and took great pride in their children's growing competence, but wished only to observe from the safe distance of the doorway. Angela Valenzuela's research on Mexicano youth and their parents in Texas suggests that the reticence of parents to enter the space of the classroom is a sign of respect for the teacher: in Mexico it would be considered interference for a parent to occupy the space of the classroom in this way (1999).

The children at Washington typified a situation of "reverse heritage"—Margaret Mead's term for times of rapid change, when children encounter and familiarize themselves with innovations before their parents do, thus reversing the usual status hierarchy between child and parent (Livingstone 2002, 124). The combination of reticent adults and eager children is typical of situations in which adults rarely work with computers at their places of employment. If such families are asked who in the household knows the most about computers, the answer is most likely to be "the children." Many parents lack the means to own or use computers and the Internet (and report feeling guilty about depriving their children of the opportunity to use new technologies).

At Washington, access to the computers is extremely limited. On the surface, the facilities seem to resemble those of Clearview, except that the computers are older, the Internet connections rarer, and the number of inoperable computers far greater. At Washington, most classrooms have two or three computer stations, but more than half the time they are not working. Because Washington is under duress for being a low-performing school, administrators often discourage teachers from taking time away from the standardized curriculum. Only a handful of teachers regularly plan computer-related projects. The computer technician is overworked and undertrained. Often she is too busy maintaining the teachers' and administrators' computers to work on those in the individual classrooms. Her expertise is in the area of IBM-compatible computers, while many of the classroom computers are Macintoshes. The position has been cut back each year to fewer hours.

A few Washington teachers have invested a great deal of time and personal expense establishing working computers with access to the district server and its software. One fifth-grade classroom has a bank of six Macintosh workstations communicating with a wireless Airport base. The teacher considers it essential to challenge her students, and encourage them in their desire to learn how to use computers. Yet often the connection fails to work, as do the computers. The teacher, with as many as thirty-five students, has little time to troubleshoot computer difficulties, as she devotes time to basic reading and math skills. She needs a "tech god" in her class to keep things going with the computers while she works with students who are further behind (Cuban 2001). But in this neighborhood, students with a wealth of computer experience are rare indeed. One of my students from a third-grade class told me that she was the only student in her room who was allowed to go on the Internet, because none of the other children had returned the required permission form. All public schools require parents to sign lengthy permission and behavior contracts regarding Internet usage: in a school with harried parents, many of whom do not speak English, this is a significant hurdle. The principal, vice principal, and teachers often complain about the parents' low level of encouragement of literacy activities. They perceive the parents as wasting money on video games and platforms when they should be buying books.

At home, only about 10 percent of the students enjoy a working computer and Internet access (citywide the average is around 35 percent). The costs are too great for the parents to bear. Many of them have purchased pre-used computers or come into possession of old Windows 95 machines, or Apple IIes, or 286s. I was frequently asked for help in repairing them. Their situation points to a fundamental misunderstanding about the digital divide. On the surface, both schools provide computer labs, technical support, and classroom workstations. The provision of the device itself is the most popular kind of solution to the digital divide. Yet both U.S. education researchers and global development scholars agree that the device is the smallest—and in many ways the easiest—part of access provision (although keeping up with the accelerating rate of planned obsolescence of computer products has nearly eradicated the benefit gained from falling hard drive prices). In order for Internet access to succeed, the device must be accompanied by a host of other expenditures for peripherals, software, and a rich network of human relationships to ensure their maintenance. With the cost of hard drives dropping, and businesses busily donating outmoded equipment at an accelerated pace for the tax write-off, why wouldn't everyone be properly equipped? Stoll calls relying on the donation of old computers a "nutty idea" and points out that the job of setting up and maintaining outdated software is truly daunting: "When old computers come without documentation, missing crucial cords to connect monitors to hard drives and hard drives to printers or keyboards, when the software is no longer available for sale, and no telephone assistance or other technical expert is available, attempting to make them work is often a pointless waste of dozens of hours of work" (1999, 172). At the Washington lab, I used a set of donated computers during the first four months of the program. The lack of standardization across machines made it very difficult for the teachers and the students. Surprisingly, the children in the class recognized the machines as donated. While they did not personally own computers, they had seen enough television to recognize an old computer when they saw one. Surrounded by secondhand things in their daily lives, they were experts at recognizing hand-me-downs, and resisted using cast-off machines.

The principal exerts constant pressure to improve test scores. Computers are a very low priority because she sees them as "not directly related to basic reading and writing." (Even the after-school newspaper class was viewed as a waste of time relative to the goal of raising test scores.) The pressure is hugely increased by the district superintendent's vow to withdraw funds from any school that fails to meet its testing targets. In this view, the Internet is just another distraction. Here is one of the conundrums of the school reform movement: the best activities for engaging student agency and investment in learning are those that are open-ended, collaborative, and involve inquiry into topics that address the students' everyday lives. But these are the first things to be cut in "back to basics" reforms like the one Washington underwent during the time I worked there. My stance is critical of educational software and Internet sites targeting children, yet it is important to emphasize how highly motivating the Internet can be for children. The pushers of educational technology are correct about one thing: children as a group get a great deal of enjoyment out of using computers. As Janet Ward Schofield and Ann Locke Davidson report in their three-year study of the Internet in schools, students consistently consider the Internet fun, deeply enjoyable, and highly motivating compared to other kinds of school activities. Provision of the Internet compels students to come to class and to arrive at class on time, to stay after school or come to school early, and to initiate projects well before their due dates (Schofield and Davidson 2002, 228). These are the kinds of behaviors I achieved in my computer class at Washington because of the Internet access: they are rarely found in other aspects of the students' school life.

The Hole in the Wall

In the mid-1990s, NIIT, an engineering research group from India that specializes in technical training began experimenting with children's ability to learn to use the Internet with no adult assistance whatsoever. The project was called "the Hole in the Wall," referring to the equipment provided: the engineers built Internet street kiosks—fortified against the punishing usage of children, as well as rain, humidity, breakage, theft, and vandalism. They unceremoniously set the kiosks down on the streets of some of India's poorest urban neighborhoods, where the researchers could be sure that there were large numbers of children who had no prior knowledge of computers or the Internet. The research team made no announcements, left no instructions, and provided no rules: they simply watched (through video and electronic surveillance) who used the machines and what they did with them. In a matter of days, the neighborhood children learned to use computers and browse the Internet. The team sardonically dubbed the experiment "Minimally Invasive Education," after the medical term for the least damaging type of surgery. The group's leader, Sugata Mitra, conceived of the kiosks as a radical pedagogy in which the only requirement for learning is the provision of a suitably interesting environment. The Hole in the Wall is technology training with "no formal instruction by a teacher or by anyone else. In fact, there is little or no intervention in the group learning process" (Mitra 2000, 2–3).

What did the children do at these holes in the wall? They clustered around the computers and taught each other how to use them through trial and error. Web browsing came easily and was a favorite activity. Kids ages five to sixteen looked up their favorite movies and music, and found free game sites to play on. The more ambitious ones figured out how to cut and paste text, and used the drawing program for pictures—even learning how to write simple words without a keyboard. While they lacked knowledge of even fundamental terms such as *mouse* or *joystick*, they gave these technical elements their own names in Hindi: for example, "sui" (needle) for the cursor, "channels" for websites, and "damru" (Shiva's drum) for the hourglass (busy) symbol. The children impressed everyone assessing the project with their ease of learning, their absorption in the task, and their ingenuity. The children developed strong attachments to the computers, and protested their removal. Most parents approved of their children's spending time at the computers, but few adults even approached the machines in the months-long installations (Mitra & Rana 2001, 5).

Why would Mitra and his organization wish to fund such an experiment if adults themselves were not the target of the project? Is theirs an act of charity—a special version for street children who never attend school? Why would they be interested in the fun children had on the computer? What would a high-end technical training outfit have at stake in accustoming children to using the Internet? What is the value of the Internet to Hindi children who do not speak English and cannot access a keyboard? What does the Hole in the Wall accomplish for NIIT?

Besides any good public relations the kiosks might generate as an act of community service, the Internet kiosks constitute a kind of blue-sky research in market expansion for NIIT, that has expanded to 43 branches, and dubbed itself "The Global IT Solutions Company." Even nonelite Indian children have pocket money at some point, and their numbers are so enormous that mass marketers are turning their attention to them. In addition, the children are being used as guinea pigs for distance education: Can the customer base be expanded to reach people without a computer, without literacy, and without any formal teaching whatsoever? NIIT is a huge operation specializing in two- to four-year IT diplomas. It works as a franchise, and graduates tens of thousands of software/IT-certified workers every year. NIIT enjoys a reputation in India for tremendous success in placing students in offshore IT jobs, as well as in transnational migrant jobs in the United States, Canada, Europe, and East Asia. Their curriculum has become a template for other schools in and outside of India. It cuts across class/caste boundaries in its reach— more so than traditional higher-education institutions, and NIITs have been established not just in the huge cities in India, but also across the country in small towns everywhere (Chakravartty 2004). Its 2005 financial report its NIIT@School business division,

> . . . that offers computer education in government and private schools recently crossed a new milestone in its initiative of taking computer proficiency to the grass root levels when it enrolled its one-millionth student from Sibsagar in Assam. Speaking on achieving this landmark, NIIT Chief Operating Officer P Rajendran said, "Enrolling the One-millionth school student in government schools is a matter of great satisfaction. It places NIIT in a unique position to be able to shape the future of millions of Indian school children in the rural hinterland." (NIIT Technologies Ltd., 2005)

Publicity materials for NIIT emphasize the achievement of "imparting a 'mass-based' flavor to computer education." (www.NIIT.com). Like new technology firms the world over, NIIT worries about market expansion in a world where the digital divide, as well as the gap between wealth and poverty, is ever widening.

There is something pathetic—or infuriating—about the Hole in the Wall experiment, and its reduction of computer use for children to a Web browser in a foreign language with no keyboard. When U.S. education scholar Mark Warschauer visited the site and conversed with area residents, parents complained that it was detracting from their children's focus on homework and study. Although mechanical failure is not mentioned in Mitra's optimistic reports, Internet access at the kiosks frequently did not work. Warschauer found that there was no organized involvement of any community organizations in helping to run the kiosk, since such involvement was neither solicited nor welcomed (2003, 143). But perhaps minimally invasive education is simply the logic of computers teacher-proofing the process of learning taken to its most ludicrous extreme.

The Hole in the Wall experiment is emblematic of themes I will return to throughout this book about children's online access in the United States: the Indian children gathered around the computer kiosk in New Delhi bear some important similarities to children in the United States whose primary access to the Internet is through community technology centers or public Internet facilities. They are not unlike the Washington students who attend my after-school class. Although as First World children they may be better clothed, fed, and educated, from the point of view of the global digital economy they share some similar features:

- × English is not their first language.
- × Their parents do not qualify for major credit cards, and therefore may not be eligible for the contracts and credit guarantees required for domestic Internet access through dial-up, cable modem, or DSL.
- × They change residences frequently, making it hard for school officials and creditors—and marketers—to track them.
- × They enjoy only limited, intermittent access to landline telephones.
- × They cannot afford a home Internet connection through a telephone or broadband, not to mention the computer necessary to stay wired.

The Washington students fall outside the target market for educational computing, and thus are left to the mass marketers of children's goods. All of these children—like those in New Delhi—can, despite their impoverished circumstances, be enticed to buy movie tickets, Coca-Cola, candy, and popular music CDs. Such purchases will most likely be made using their own pocket money rather than their parents'. At the bottom of the market, parents are not the target, and there is little valuable marketing information to be culled about the adult household members from monitoring their children's Web activities.

On first glance, it appears that the Internet is everywhere in children's lives—from the streets of New Delhi to the suburbs of southern California. On closer look, however, it becomes clear that the children of elites and urban professionals experience new technologies in a qualitatively different way from poor children. Far from being a leveler of class differences by opening access to information, adoption and dispersal of the Internet has deepened social divisions along the lines of class, race, and ethnicity, and it has done so both within countries and among countries. In the case of intracountry disparity, state policy helps to explain the differences, but it has also been determined by the fact that English is the dominant language of the Internet. Even among developed European nations there exist large disparities due partially to language: Finland has far more Internet access than France, for example, and it also is a country where English is considered a necessity.

Staggering inequalities exist within countries, such as India, where there are hundreds of millions of people living in poverty, on the one hand, and a highly developed technological sphere on the other. Across the globe, Internet access favors those who are native English speakers and live in rich and technologically developed nations. But fluency and residency are not enough: while high-income U.S. households approach the saturation point with Internet access, low-income groups, and Blacks and Latinos, regardless of income or education, lag far behind. Employment figures echo the demographics of the digital divide. In Silicon Valley, Latinos make up 23% of the population but hold only 7% of the high tech jobs, and none of the management positions (*The Economist* 1999). Nationwide, Blacks make up only 5% of all programmer positions; only 6% of all technology jobs (Keefe 2002).

The digital divide falls out across familiar lines of First and Third World, West and East, developed and underdeveloped countries. In the same countries where transnational corporations exploit workers, parents and children of the richer classes purchase the same products of a global children's consumer culture, from Barbie to Harry Potter to Pokemon. While per-capita spending by children varies across regions, the reach of the children's culture industry is truly global, and the Internet is part of its marketing strategy. In the many countries where children are exploited as factory workers in export processing zones, there also exist classes of children who consume the products of this global market and enjoy Internet access from home.

According to a recent study by the Annie E. Casey Foundation, significant differences in access persist in the United States among children of different classes and races. Of children living in households with incomes over $75,000, 95 percent have computers at home and 63 percent have Internet access. Sliding to the other end of the spectrum, of children in households with incomes less than $15,000, 33 percent have computer access and 14 percent have Internet access. That means that only a third as many poor children as affluent children enjoy these privileges, despite overall large increases between 1993 and 2001. (The study also found that the lower-income students were more likely to use home computers for games than for homework and standard software applications such as Excel, Windows, and so on.) It has come increasingly clear that the struggling working and middle classes simply cannot afford the investment (Tomsho 2002).

As computer use moves increasingly to a model where the Internet is used to serve up the necessary software and databases on an as-needed basis, families with children may find that the home Internet connection itself is more important as an enabling feature than the type of hardware one owns. For the working-class children who attend my computer lab, possession of a high-speed Internet line from home is virtually an impossible dream. Even if Internet appliances replace computers and lower the costs of purchase, contracts with phone companies for DSL lines or cable companies for high-speed cable service are priced far beyond the means of these families (and there is no plan to wire their neighborhoods for the cable services due to the low-income demographics). These parents can only intermittently afford to keep a single telephone landline in use, or to pay the TV cable bill. The obstacles are threefold: tight household budgets, difficulty in negotiating with the utility providers when English is a second language, and a high rate of transience, where moving among apartments to beat the bill collectors, or moving in with friends or relatives during periods of unemployment or debt, are commonplace.

Federal support for computer access for poor children and children of color is dwindling, a phenomenon referred to as "digital red-lining" (Tomsho 2002). President George W. Bush drastically cut the budget for community technology centers shortly after taking office, and has issued a series of proclamations that the digital divide was nothing but a myth—this despite the fact that Internet connections in wealthy schools outnumber those in high-poverty schools by a ratio of two to one (*Wall Street Journal* 2001). The Bush administration abandoned the Clinton-Gore position on new technologies. The Federal Communications Commission chair Michael Powell compared the digital divide to "a Mercedes divide—I'd like to have one, I can't afford one" (Bridis 2001). The State of California paid skyrocketing prices for energy to Enron and its subsidiaries before scandal and bankruptcy broke out: the result is a devastating cut in school budgets and technology initiatives. The bust of the Internet economy means that philanthropic efforts to bridge the digital divide have evaporated.

At the top of the children's market, many marketers hope to reach parents' wallets through their children. Hardware and software companies have promoted the idea that children—with their instinctive attraction to computers and their admirable lack of technophobia—would be the gateway into homes. Children could convince parents to buy the hardware (and update it frequently so it would be suitable for the latest technical advances in gaming), subscribe to Internet service providers, and buy an extra telephone line or a broadband connection. These kids have connections at home as well as at school, and many are among the 14 percent of U.S. residents with DSL lines at home. In contrast to the low end of access, 67.3 percent of Americans making $50,000 to $75,000 per year and 78.9 percent of people making over $75,000 per year use the Internet. Even during a time of stalled growth, the over-$75,000 group saw a 34 percent increase in the share of households with home Internet access between August 2000 and September 2001. Children in the United States are five to six times more likely to have Internet access at home than their European peers; they are ten times more likely to have a computer in their bedroom (Livingstone 2002, 165).

The largest gaps between rich and poor children will be apparent in the arena of multimedia production: the very video, audio, and gaming applications most popular with children. "[T]he economics of the information technology industry, together with the social stratification of educational systems, mean that multimedia creation is highly *inaccessible* to the masses. On the one hand, while the cost of computers and Internet access continues to fall, the cost of the hardware, software, and bandwidth necessary to create the newest forms of multimedia will always be more expensive" (Warschauer 2003, 202). In other words, working-class children have little chance of enjoying the kind of computer and Internet access that is residential and high speed, the kind that facilitates music downloading, online gaming, and instant messaging. And while these activities seem like nothing more than play, we know that they are vital to social inclusion. This is what is at stake in children's computer use—far more than simply learning keyboarding or how to save and delete files. As Warschauer points out:

> in both wealthy and poor countries, the singular concentration on the computing device itself, to the exclusion of other factors, is a shortcoming of many well-intentioned social programs involving technology. . . . What is at stake is not access to ICT [information and communication technologies] in the narrow sense of having a computer on the premises, but rather in a much wider sense of being able to use ICT for personally or socially meaningful ends. (2003, 12)

An increasingly important dimension of this inclusion is entertainment. As I will argue in this book, children look to computers for entertainment—from video game characters to wrestling matches—and they find fan activities deeply compelling.

There have been unintended consequences of equipping our educational institutions with computers and making new technology the top spending priority at many institutions, from kindergarten classrooms to major universities. As Sonia Livingstone points out, in her study of families' use of new media, the distinctions between these two realms are increasingly blurred:

> the locus of considerable uncertainty, is the part played by the arrival of the domestic computer and by the discourses, institutions and ideals of education. The institutional aspect of this inter-penetration of entertainment by educational goals should not be underplayed, for although leisure has traditionally been relatively private, part of the lifeworld, education by contrast has long been central to the project of the state, a center-piece to the production of a future competitively skilled workforce. Overwhelmingly, the case for young

people gaining access to ICT at home or at school is presented across public and policy as an educational one. (2002, 230)

Even if the costs of updating hardware and software were not barriers, in situations such as in the Washington lab, where hardware has been generously supplied by grant awards, other kind of resources are demanded. These include social networks of people who have knowledge of computers and can provide assistance, and a level of literacy that will provide the ability to type, read manuals, use e-mail fluently, and benefit from written exchanges online. There is also the need for the incentive of relevant content, because the Internet is so heavily skewed toward white, English-speaking professionals who are interested in making purchases online. There is a "multiplying factor for social inclusion" of the groups who successfully gain access. Warschauer has emphasized that the necessary resources—physical (the box), digital (the connection), human (literacy), and social (friends and family members who are also online) for Internet access are "iterative," mutually reinforcing:

> The presence of these resources helps ensure that ICTs can be well used and exploited. On the other hand each of these resources is a result of effective use of ICTs. In other words, by using ICTs well, we can help extend and promote these resources. If handled well, these resources can thus serve as a virtual circle that promotes social development and inclusion. If handled poorly, these elements can serve as a vicious cycle of underdevelopment and exclusion. (2003, 24)

Since the collapse of the dot com economy and the rapid move by U.S. corporations to outsource work involving computers to East Asia, the promise of employment for U.S. students with computer skills in jobs that would pay a middle-class salary with benefits has dimmed. The jobs that the high-tech sector provides are both fewer in number and far less desirable than many imagined in the 1990s: as we move into a period of severe state budget cuts in school spending, and deepening unemployment, it is time to take a long, hard look at the benefits and the costs of sending children online. What is the point? Is the experience worth the costs? What kinds of teaching and critical thinking skills do children need to navigate the Internet?

chapter 2

Children, Politics, and the Internet

Stories from the Journalism Classroom

Remember when the Internet promised to cure a host of social evils? I left the college classroom to begin teaching children ages eight to twelve, at that now distant moment when the economy was booming, and President Clinton vigorously supported the construction and improvement of school, library, and community computer labs with Internet connections as a means to democratize computer access for children. In the late 1990s, tremendous public concern existed for those children being left behind by the information age and there was enormous faith in the power of the Internet to remedy educational inequity.

I was a believer. Like many other media educators, I advocated Internet use by children to help bridge the digital divide, to revitalize primary education by broadening the curriculum, and to ameliorate the gender gap in skilled computer work. Having taught at major research universities throughout the 1980s and 1990s, I was pushed—along with the rest of the faculty—to adopt all manner of digital communications early on and in as many capacities as possible. New technologies were freely, readily, and rapidly becoming more available at universities. As a film- and video maker, I found digital media and the capability of editing images on computers miraculous. In graduate school, I had hand-spliced 16mm film with heated cement; now with a single stroke of the keyboard I could edit, re-edit, and render fabulous special effects in a hundredth of the time it would have taken me to do it the old-fashioned way. I was using e-mail on a daily basis by 1988— spurred to do so by the capacity to exchange files with colleagues in Europe, with whom I was collaborating on a book. E-mail eliminated weeks of waiting for drafts to arrive via overseas mail, drastically reducing production time on our book. These experiences drove me to master new technologies as they appeared—it seemed necessary to keep pace with the undergraduates, who showed up every year more knowledgeable about software we had yet to introduce in our college courses. Strenuously encouraged by university administrators to bring new technologies into the classroom, I worked hard to introduce students to new hardware and software whenever and wherever I could. In retrospect, it is important to consider that at all levels of teaching, the take-up of new technologies was a top-down mandate, all but required of teachers. I produced a CD-ROM for teachers on a media education topic and distributed it for free for classroom use. When, in 1999, community leaders in a working-class neighborhood invited me to design a computer lab and an after-school program for children, I jumped at the chance to escape the ivory tower. A coalition of community groups had banded together to build a school annex to relieve overcrowding. They had included in the plans a spacious "multipurpose room" with a T-1 line (enabling high-speed Internet access from banks of computers). At the time, the only public Internet access in the neighborhood was a single computer terminal at the local library nearly a mile away. The program allowed me to put to good use my own knowledge of pedagogy and computers to benefit children who never dreamed of attending the universities where I had received my training. By helping children develop a sustained interest in computer applications and Internet usage, I hoped the kids might see more benefits of education, and begin to imagine themselves as future university students. The course met twice a week, for ninety-minute sessions, and I was assisted by six undergraduate tutors. (The course has continued in a slightly different format.)

The goal of the program was to develop students' skills as communicators representing their own communities. The Internet and computer software would be made more relevant to students by using them to report on the students' immediate community and their creative decisions representing their own lives and their own neighborhood. The newspaper class was designed to integrate Internet research more fully into student writing and to link the computers to more intellectually challenging work than is possible in more routine language arts curricula, especially those closely tied to assessment through standardized testing.

The newspaper class was planned to follow a child-centered pedagogy in which the students would generate their own topics. In *Free to Write*, teacher and journalist Roy Peter Clark (1987) advocated teaching journalistic skills to children as a way to enhance their interest in reading and writing. According to Clark, published journalism has significant advantages over other kinds of writing because it provides an audience for students' writing and allows teachers to value children from the moment they enter the classroom as knowledgeable sources of information about their communities. Furthermore, this kind of writing teaches the value of reflecting on life's events through the process of recording what happened, why, and how people feel about it. The power of literacy is demonstrated vividly for my students, who, for days after the paper comes out, feel "famous" in their school and their neighborhood.

Educational theorist and activist Paolo Freire's concept of critical consciousness shaped the design of the newspaper project. Freire (1994) emphasizes equality and dialogue between teachers and students, recognizing the many diverse forms of knowledge that are already present in every classroom, although rarely acknowledged in traditional pedagogy. In Freire's view, literacy should be directly tied to the validation and dissemination of these different forms of experiential knowledge. Freire argues that intellectuals can benefit from receiving knowledge from their students, and questioned the strictly limited forms of knowledge valued in the academy and by traditional status hierarchies: education becomes a transformative experience for both teachers and students.

Critical education theorist Michael Apple (1996) has adapted Freire's work to the context of U.S. public education, and Jim Cummins and Dennis Sayers (1997), have applied Freire specifically to the use of new technologies. In the design of my after-school class, the linking of journalism with new technologies came from a desire to use news writing and photojournalism as a forum for discussing public issues, providing political education, and exploring the potential for children to take on activist roles in their communities. Apple has written about the need to combine "the 'practical' with the 'critical' and 'theoretical'" at all educational levels: "This is especially the case for those students whose life on the margins is produced by the decisions of economically dominant groups, and then legitimated by the discourse of neoconservatives whose own vision of social justice seems to amount to no more than blaming the victim" (1996, 102).

Throughout the project, children are encouraged to think of writing as something meaningful and shared. In an adaptation of Freire to the environment of new technologies, the goal is to encourage children to conceive of the Internet as more than a game machine or a new delivery system for music, television, and film, and entertainment-industry publicity. My hope was to encourage children to use the Internet as a means of voicing their nascent critical consciousness. That hope has dwindled, not only because the Internet's many distractions related to the entertainment industry are deeply compelling to children, but because Internet research has proven truly daunting for these elementary-school children. Here I will discuss three examples of the ways children in my class have generated—and discarded—story ideas of political relevance; how their stories reflect their conflicting and overlapping age, class, gender, and ethnic identities, and how their use of the Internet has challenged me to reexamine the project's goals and the possibilities for children to use the Internet to voice their political views.

Finding Stories

When we first asked the children at Washington, "What interests you in your neighborhood that you would like to write about?" they were silent. Only after months of reflection and observation did I begin to understand their reticence. First of all, the students were largely unfamiliar with newspapers as a genre and a form of communication. When we passed out sample newspapers to teach about different sections and types of stories, sports, entertainment, and comics were by far the most familiar types of news. Rarely could students make the leap from a published story they read to an idea for a story of their own.

A second factor appears to be that the students, accustomed to more authoritarian teaching methods, were unfamiliar with the middle-class model of family and school life in which self-expression is prized and children's ideas and interests are taken seriously. Child-centered pedagogy—so prevalent in the upper-middle-class professional milieu of Clearview in the suburbs—was a new experience for these children. In fact, teaching methods had become more authoritarian as the school district was undergoing radical back-to-basics school reforms. (As the class has progressed, and created its own local reputation as a desirable place to be after school, with friendly and trustworthy teachers, the children have become freer and less restrained in their venturing of new ideas.)

Finally, the Washington students had no experience of the kinds of "safe" topics that typically fill high school newspapers, school newsletters, and kids' pages: music, art and dance lessons, sports teams, pets, and hobbies were out of financial reach for most of these families and beyond the realm of the students' experience. At the companion project I taught at Clearview, there was no lack of story ideas, most of them closely tied to public relations accounts of extracurricular activities. The students' lack of familiarity with the "happy news" that fills most community newspapers had its advantages as well. When I suggested topics that were school-related, the children often rejected them with the comment, "That's boring." They were not interested in teacher profiles, or interviews with school administrators. Instead they usually chose to write about school only in terms of "extracurricular" topics: the poor quality of cafeteria lunches, teachers' activities during recess, or popular games on the playground.

Leads for stories usually took the form of children's observations of their environment or of adults close to them. Newsworthiness for the students often correlated with the darker side of world events: violence, fraud, cheating, and injustice. The students were especially eager to talk about the September 11 terrorist attack and the disputed returns of the presidential election of 2000. Violent crime and kidnappings, of which several notorious local cases took place during that year, captured their attention: school shootings, child kidnappings, and murders were topics of great fascination, and students independently researched these. In the case of kidnappings, students printed out news reports on these topics and made posters soliciting information about the abducted children's whereabouts.

The students' perspective on safe topics of national interest—one radically different from their more affluent Anglo peers at Clearview—was made vividly clear during a series of presentations on the Winter Olympics by guest speakers from the medical school. The speakers were a team of outreach workers from Community Pediatrics who had a dual focus on encouraging exercise and improving nutrition. They presented to our class a well-prepared lesson and set of activities that had regularly been presented at more affluent school sites. The speakers began their presentation by asking students how many of them were watching the Olympics: not a single child raised his hand. (Frequently the families in this neighborhood cannot afford cable television, leaving them with only a few channels and poor reception.) When the teachers asked if anyone in the class had ever participated in winter sports, again no one responded. Anxiously, they asked if students could name any of the sports played in the Winter Olympics, and a few students answered but could name only hockey and ice skating. In desperation, the teachers asked if anyone could name any Olympic athletes or anything about them. As soon as one student mentioned athletic disqualification for flunking the drug test, lots of hands went up and everyone in the class became quite animated. This was the one angle on the Olympics—drugs and drug testing—that most of the children were familiar with and could take an interest in. This might have been a teachable moment: the community is one in which drug problems are common, and some of the students have family members in drug rehabilitation or in Alcoholics or Narcotics Anonymous, and have experienced dramatic changes in their lives because of adult drug use. Participation in team sports, soccer moms, and fitness centers, in contrast, are virtually nonexistent in the community. The speakers were dismayed, however, because the children's fascination with drug use among athletes made their planned lesson about Olympic endeavor as heroic and patriotic difficult to continue. "Can you think of any *positive* examples of something you have heard about the Olympics?" Silence fell over the students again.

Other guest speakers generated questions that were clearly inflected by the students' social positions, and demonstrated the kinds of perspectives Freire valued in his pedagogical model. From the very beginning, students proved to be surprisingly adept at interviewing as a method. The writing of interview questions came very easily to them, and their verbal exchanges with adult visitors produced lengthy, lively transcripts. The students enjoyed handling the tape recorder, hearing their own recorded voices played back, and watching the tape transcribed onto the computer screen. In a class exercise repeated on a regular basis, students interviewed each other or the class assistants, and were directed to think of questions that could produce long, more narrative replies—to go beyond a yes or no answer. This lesson stood out vividly in the students' minds: they retained this particular piece of advice about reporting technique better than any other. The etiquette of interviewing was also discussed in class: one of the first questions on the children's minds was how much adults earn in their jobs. After some of our interview subjects suffered noticeable embarrassment, we discussed why income can be a sensitive topic and how to decide whether it is too personal a question to bring up. The children's interest in earning power indicates their awareness of class issues: seen from the light of their critical consciousness, income shouldn't be a "personal" question, but a political one.

The students' questions often steered guest speakers toward matters of identity. Students' interests reflect their attempts to negotiate a world deeply inflected by racial and ethnic concerns. When a Hindi documentary producer spoke to the class about her work and her native country, the students focused on how many different languages she could speak and used this to confirm a sense that being bilingual could help a person get a good job. When a music critic spoke to the class about the origins of popular music, their bottled-up interest in and confusion about Michael Jackson (and his repeated cosmetic surgeries) came pouring out. The entire session was dominated by questions about whether it was possible to change one's race, and whether Jackson should now be considered white. A lively argument followed among the students, who were eager to speak on this topic, and exhibited a wide variety of understandings of racial categories. As Audrey Thompson has argued, the "colorblind" model imposed by so many teachers effectively shuts down discussion of a topic—race and racism—that students are struggling with on a daily basis (1998, 520). It is a disservice to students to exclude from classroom discussion issues of class and race that they are negotiating throughout their everyday lives. Instead whiteness is embedded in the "colorblindness" discourse, which is "universally framed and has thus sidestepped the issues of racial imbalance implicit in colorblindness" (Thompson 1998, 524). In another class session, when visiting filmmaker Zeinabu Irene Davis showed a dramatic film about slavery (*Mother of the River*, 1991) that included scenes of a brutal whipping, the children turned their attention to the details of staging this scene, how it had been made to look real, and what pain, if any, the actors felt.

The visual world, rather than verbal discussion, often provided the most successful basis for generating topics. Taking pictures was a better way to instigate discussion of community issues than sitting around in the classroom brainstorming for topics. The students always saw much more in their environment than their teachers noticed. In one class routine, students were given lessons in photography and went out in groups of four or five with an instructor to walk through the neighborhood and take pictures of anything interesting that might lead to a topic for a news story. This continues to be the favorite activity of the students. By emphasizing visual literacy as well as verbal literacy, it is possible to demonstrate how much children already know about their community and provide more immediate gratification, in the form of digital picture taking, than comes with news writing, which is an admittedly long and laborious project for some children. If being a writer is a distant, vague, or daunting prospect for these children—just plain hard for them to imagine—being a photographer or a photojournalist is immediately embraced, avidly adopted. Holding a camera and being able to take pictures at will was immensely empowering for them. At each opportunity, students clamored to be first to take a turn with the camera, whether they were photographing a subject related to their own story idea or not.

When we took the cameras out into the streets, large lists of topics were generated right away, and these touched on some of the community's core issues: a trip to the local bakery led to a story about parents working night shifts (headline: "6 to 6 Helps Working Parents"), graffiti led to discussion of gangs (headline: "Graffiti Artist Quits His Old Ways"), the presence of squad cars or police was linked to the citing of skateboarders for traffic violations (headline: "Is Skateboarding a Crime?"). The children felt a special identification with the dispossessed. A recurring theme in their story ideas was displacement: What happened to people when their apartments or homes were torn down—where did they go? This came up in terms of the housing that had been removed to build the new school annex in which we were located. Questions of history—why is our neighborhood the way it is—were most likely to come up through the physical environment. How long had a building been in place? What different uses did it serve? What was here before? Urban planning—a topic rarely introduced in any elementary school curriculum—was one of the subjects of greatest interest and debate among the children.

The first edition of the paper (pre-Internet) was comprised entirely of local stories generated through the photography sessions: the recreation center's boy's basketball team, the nearby pet store, a local magic store owned and operated by a new-age group affiliated with Wicca, a group of local blues musicians, graffiti, and the local fire station (see illustrations 1, 2). Most stories were ones for which interview subjects were readily available and could come to class—the pet store workers, the firefighters, and the basketball coach.

When Internet access became available three months into the beginning of the class, it presumably became possible for students to research stories of local relevance online and to conduct interviews with knowledgeable subjects, experts, and authorities via e-mail. The capacity to look up anything and everything on the Internet (or so we thought) coincided with the construction of ambitious stories that had the potential to mobilize the students around local political issues.

Illustration 1. Before Internet access, children found story topics in their immediate neighborhood, like graffiti.

Illustration 2. The Magic Store located on the main business street fascinated the children.

The Ballpark

During a brainstorming session in the fall of 2000, a ten-year-old Latino boy I will call Andres mentioned proudly that his stepfather was employed at the construction site in downtown San Diego where a new baseball stadium was being built. He was worried, because construction had come to a stop, and many workers had been laid off. So far, his stepfather was still employed. Three more boys (Diego, Americo, and Reyes) latched on to this topic as an interesting and controversial one, and joined Andres in working on the story. From the beginning, it was clear that their interest in the story did not stem predominantly from fan feelings for the Padres—the high price of tickets to major sporting events ruled out their attendance, and they seemed to have little cultural identification with the local teams. (Wrestling is much more likely to be the sport of choice for fathers viewing with children in this community; it is discussed as a topic charged with issues of race and ethnicity in Chapter 5.)

What interested them most was the ballpark as a massive construction site, their curiosity about what happened to those whose homes were torn down for the construction, and their sense of the city's work stoppage as unusual, confusing, and maddening. In his first draft, Andres wrote, in an editorial tone, "Let's donate money to the ball park construction project. They need to hire more people to build the ball park. The city needs more money to continue the project." Here Andres presents the ballpark development as a kind of charitable endeavor, requiring individual donations. His position on the construction stoppage is fueled by anxiety over his stepfather's potential unemployment.

The group of boys formulated a series of questions that would guide them in their efforts to gather information on the Internet:

- × Why did they run out of money?
- × What happened to the houses and the people who lived there?
- × How many buildings were destroyed? How many people were in the buildings?
- × Why did they start building and then stop?
- × What will the finished ballpark look like?

When the boys began to search the Web for answers to their questions, first searches under the keywords *Padres*, *baseball*, and *stadium* led to dozens of ticket offers, season ticket solicitations, stadium seating charts, and sweepstakes for winning tickets. Routinely, children encounter a barrage of marketing devices, opportunities to make purchases, games of chance, and sweepstakes at all of the websites that interest them; this project proved no exception. The boys had been instructed on the use of keywords in Internet searches, and knew the name of the online site for the local newspaper, the *San Diego Union-Tribune*. After adult assistance in adding some keywords to the search, each of the students wound up at the official Padres website—hardly an unbiased source of news. They found some legitimate journalistic pieces, which seemed to be reprinted on the site because they were favorable to the Padres' position that the city should pay for as much of the construction as possible and by whatever means necessary.

Reyes and Americo were delighted to find as many as ten articles on their topic, and busily printed them out. Not one of them suspected that stories printed on the Padres' official website could be slanted toward one side of the ballpark story. The boys were very satisfied with the extensive timeline provided by the Padres-specific website. Indeed this polished, highly truncated version of local events seemed to answer all of their questions. The timeline appeared on a page sponsored by a local Indian casino and a local construction company. To their minds, their work was done—key dates from 1996 to 2000 were outlined, stressing the inevitability and the accomplishment and cheery news of the Padres. In this, they are not so different from many adult Internet users (and certainly many university students), who are often unaware of the distinction between sponsored content and news content on commercial websites. Unfortunately, the timeline provided by the Padres excised from the story all concerns raised by the opposition: that the ballpark was financially foolhardy; that public dollars were subsidizing a private commercial concern; that it represented a "big box" strategy for urban planning, in which the problem of a blighted area that would take years to develop is solved by simply tearing everything down and putting a huge, expensive, but relatively low-use operation in its place; that the ballpark project had succeeded only through illegal lobbying of the mayor and city council members (one of whom was later criminally charged); that the ballpark constituted a massive investment in tourism and service industries at a time when the San Diego economy needed to diversify and provide more living-wage jobs. Multiple lawsuits had been filed against the city attempting to stop development and construction of the ballpark, but information about these was absent from the websites generated by the boys' searches.

It was much more difficult to locate information on the Internet about opposition to the ballpark, and this lack of the other side's view became another obstacle for the boys to overcome. In an attempt to build some quotations and more detail into the article, and perhaps capture more of the controversy, the boys e-mailed the mayor's office to get direct answers to their questions, especially about the houses that had been torn down. We e-mailed the mayor and the mayor's director of publicity with the list of questions and a description of the newspaper project.

The kind of e-mail exchange the students undertook is frequently hailed by Internet enthusiasts as one of the most educationally stimulating aspect of classroom Internet usage. In theory, children who do not have access to the phone (or the skills to get their calls answered), or the transportation to travel to a location, can interact with knowledgeable experts online and thus greatly enhance their research, understanding, and writing. A variety of projects have been developed linking students across national borders, or across economic and social boundaries, to facilitate language learning, or understanding between blind and sighted students, or among different religious and cultural groups. Such projects have been based on the belief that any kind of exchange with others who are in some way different will enlarge student's perspectives and worldviews. Schofield and Davidson argue that here, too, the outcome depends very much on the teacher's "crucial role in shaping the kinds of experiences students have in such exchanges and in influencing if and how they will broaden students' perspectives" (2002, 250).

Rather than educational projects involving peer e-mail exchanges, the newspaper project encouraged students to use email to elicit information from authoritative knowers. The disappointment rate was very high for the Washington students. The problem stems from one of the unanticipated outcomes of Internet technology: for adults working online, the sheer amount of e-mail received means that it has become increasingly difficult to answer individual messages. The chances of the students receiving an answer are relatively slight, unless the person is forewarned, and unless the teacher uses personal influence and connections in order to guarantee an answer. As Schofield and Davidson (2002) report, attempts to communicate with eyewitnesses and outsiders who could furnish information were often rejected or ignored. E-mail is a discouraging literacy activity, when students send off messages to adults that will never be answered and that simply disappear into the ethernet. Waiting for more information to arrive before writing the ballpark story, Andres and the other boys checked the e-mail account for the class each class session for three weeks, before giving up. By that time, Andres' stepfather had been laid off, and interest in the topic was flagging.

In the absence of a broader context for the ballpark and its construction, the boys focused on the appearance of the park. Searching on the topic of the size of the park, yielded only boosterism: "Big Crowd, Big Win," "King-Sized Win," "Home, Sweet Home." One brief, early draft exemplifies the confusion the boys experienced in their research attempts. Why would we need a new ballpark if it isn't going to be larger than the old one? The boys wrote: "The new ball park will be smaller than the Qualcomm stadium. Many persons don't want the new stadium. They say that it is an embarrassment to San Diego because of the construction stopping. It can only hold 42,000 people." While they were able to get the number of seats from their online investigation, they were not able to locate any explanation for the curious fact that the ballpark would be smaller than the existing one, or any explanations for why some citizens did not want the ballpark—except that it is embarrassing that it stopped. The architectural drawings of what the projected ballpark would look like also attracted their attention. One student wrote "It's pretty cool and you don't have to buy tickets but you can see from the outside," expressing, probably with great accuracy, that he was likely to come into contact with the ballpark not as a patron but as an observer of the construction and an outsider looking in. (These free spaces for viewing the ballpark were ultimately eliminated once construction was completed.)

When a tutor provided the boys with some AP wire stories with more substantial analysis of the controversy (under titles such as "San Diego Mayor Is Accused of Illegally Angling for Stadium"), the boys declined to read them. Instead they gravitated toward the less informative but more exciting articles on the Padres' website, notices from *Sporting News* with four or five action-packed photographs per page. The Padres' marketing efforts directed at the Latino community paid off here: the boys were pleased to find headlines such as "Batazos inaugurales," "Que dia bonito," and "Monterrey fue el escenario y Padres de San Diego el embajador del beisbol de Grandes Ligas."

Frustrated with the effort required to cull through the fine print in the AP stories, and discouraged by the lack of response from the mayor's office, they salvaged the story by doing "vox pop" interviews based on their questions. The story was written as a public opinion story, under the headline "What Happened with the Ballpark?" The final version of the story displays the students' original concern for the workers being laid off, and the situation in which workers are given conflicting or incomplete information about the financial status of their employers. The topic of the displaced homeowners was impossible for them to pursue on the Internet, because the story was being actively suppressed by the local newspaper, the city government, and the Padres owners. As Ben Scott contends, public journalism has worsened in the age of the Internet: "What's left out is what's always been left out, new and public information affecting the lower half of the socioeconomic spectrum. Online news content is very often market-based journalism of the rankest order. . . . More is not better when it's more of the same" (forthcoming, 22). There was information to be found on this topic on the Internet—one class action suit was filed on behalf of the displaced residents, and one of the critical objections to the ballpark was that it would reduce even further the amount of affordable housing downtown—but it was highly unlikely that the average student would have come across this information online, as sponsored websites top the list of search engine results.

The Internet was supposed to greatly diversify the sources of information and points of view available to student researchers. But it takes a great deal of sophistication to judge online information. The reading skills involved in deciphering the small print characteristic of most browsers is daunting for elementary-school children. Advanced critical thinking skills are necessary to sort out good and bad information on the Web. And fundamental knowledge of the business of Web publishing would have been required for Andres and the others to understand why the Padres official site would appear first on the lists provided by search engines. Children need to know, from their first attempts to research online, that space has been sold to the highest bidder, and that the hits that appear first on their searches have been paid for.

School Lunch

The ballpark construction story was an ambitious undertaking, following relatively fast-moving current events that were somewhat beyond the experience of the students. A more familiar and presumably easier-to-research topic that was a favorite among students was food. While food is usually framed as a neutral, non-political topic, the students used food to express their resistance to adult norms, and their resentment of the institution of school itself.

In the newspaper class, eating was the favorite activity, and slowly this avid interest was captured for the purposes of journalism. Food was always talked about, as we started each class with snacks at tables set apart from the computers. Because the class met after school, the students were hungry. The school is one in which 90 percent of the students qualify for free school lunches, so the cafeteria and its offerings were one of the first things that came to mind for a story idea. Some students brought additional snacks to class—sometimes from the 7–11 store on the corner—and these were always the focus of attention and controversy. At the two-year mark, when we interviewed the children in pairs about the class for the purposes of evaluation, the overwhelming majority named the snacks—not the computers or the Internet or the teachers—as the best thing about the class. This was a humbling finding, but an important indication of the powerful role of food in the children's lives.

For six months, our class was visited by public health educators who devised many imaginative, hands-on exercises for the children to teach them about fat and sodium content, nutritional requirements, and fast food. Considerable effort went into teaching the children how to read labels and directing them to the wide variety of websites provided by food manufacturers, dairy associations, and so on for information. Some of the most elaborate websites, of course, are those of the candy manufacturers. The students wrote a couple of very good articles comparing the nutritional value of the classes' two favorite snack items (Flamin' Hot Cheetos and Sour Patch Sour Apple Straws), on the one hand, and popular fast-food choices on the other.

In this case, Internet research went smoothly—the students became very adept at finding nutrition charts online, searching for specific information such as fat content per serving, and writing it down. Of course, it was much easier to find this kind of information for commercially processed, branded food items, like Cheetos, than it was for something like an orange or a piece of celery. Locating information on the home pages of food corporations means making it past the elaborate eye-catching games that fill the pages of candy, cookie, and cereal makers. This is another instance in which the Internet veers toward the commercial world—in the case of food, the content and style of presentation are instantly familiar to the children from television. After instruction and monitoring of their browsing, it became possible to use these commercial sites to gather data about nutrition. Although the Internet can serve as a valuable resource in this way, it is easier and less distracting to read the same information off the label.

Much more problematic than the stories about nutrition was the desire to write about school lunches. At first the children were interested in capturing lots of quotes from other students describing how terrible the food was. When searching the Internet to try to find the school districts' menus for the month, hundreds of sites came up with suggestions for recipes for packing nutritious sack lunches, and the geographic reach was very wide. Attempting to locate specific information about a school with a common name (such as a president's) is truly daunting. Repeatedly, students would visit dozens of websites for schools of the same name in distant states, from Arkansas to New Hampshire, and often they found it difficult to recognize that this was not their own school.

As the students got closer to conventional reporting methods, censorship became overt. School lunches are a very controversial topic in California: increasingly, legislators and school boards are attempting to sever contracts with soft-drink and fast-food providers, for example. Originally, the students assumed they would be able to photograph the cafeteria and interview the lunch workers. When the students asked to interview the "lunch ladies," they were first stalled, then eventually told that the food service workers' supervisor had forbidden the interview. The only person we could get to agree to be interviewed for the story was one mother who had formerly worked in the cafeteria. Instead, the children were directed to the vice principal, who gave a thoughtful but deliberately neutral account of the difficulties of serving lunches. In the final story the concluding paragraph read:

We asked Mrs. G., the vice principal, why the food tastes so bad. Mrs. G. explained, "Well, you know food tastes better when it is home cooked and you are only cooking for a few people, when you have to make it for so many people at one time, it loses its taste. It will never be like a home cooked meal. A delivery company called Food Services delivers the food to [Washington] and many other schools. Sometimes they have problems at the cafeteria if they run out of food for that day. Mrs. Betty is in charge of the school lunches. Remember, many school lunches are nutritious—you might not think so but they're still good.

The concluding sentence is surprising (since the students originally proposed that the headline for the story be "School Lunches Suck") and represents the students' resigned acceptance of the party line after their curiosity had been answered with public relations jingoism.

Blueprint for the Schools: The Story That Got Away

One spring day in 2000, three of the oldest, most popular, and most vocal boys in the class ran in the classroom door and excitedly requested that they be allowed to perform a rap for the class during snack time. I agreed, unsure of what was going to happen, but pleased to see them so excited about something and curious about what they would do. Three twelve-year-old boys, one Anglo and two Latino, hovered in a corner writing out the rap, dividing up song lines and rehearsing.

The rap was a protest song they had created in response to the news that a reform was going to be instituted in the district that would cut six hundred teacher's aides from the payroll in order to add more classroom teachers, more training for teachers, and remedial summer school classes for students. The Blueprint was mandated by San Diego school superintendent Alan Bersin, who came to his work in school administration from a career as a district attorney and who had national political ambitions. The Blueprint was designed by Manuel Alvarado after his controversial tenure in the New York public school system, and was vocally opposed by the teachers union, by parents, and by students.

The Blueprint Rap
This is a rap telling about this ugly crap
After I'm done with this, gimme some dap
This Blueprint is in the gap

This is a story about a guy
And his story is up in the sky
But all we can do is sigh
So Why, Why, Why?

This is the right way to tell ya
The Blueprint will take away

All our supplies and extra classes
Blueprint. Na, Na, Na.
Why you do that?
I'm trying to be a good kid
But you keep taking my stuff
I pledge to you I will do,
That I will do stuff
Blueprint. Na, Na, Na

Why do you want us to go to summer school?
Because we don't want to waste
Our summer vacation
On the stupid Blueprint
Thank you, thank you, thank you very much.

Blueprint, I have some stuff to tell you.
You don't have the right to do that.
Now know your role and give it all back.

There is a striking sense of outrage here, expressed with power and poetry. The students, who struggle to articulate their social positions, were motivated in these exceptional circumstances to find some means of public expression. Being at the center of a stirring local news event inspired them to object to their position in the school system, as one down, as always criticized, as not smart. The boys heard the rhetoric of "low-performing schools" and were able to label it what most critics agreed it was: a punitive measurement targeting low-income communities and bilingual students (Berliner and Biddle 1995; McNeil 2000). Their greatest dread was mandatory summer school and the stigma of remedial courses. But they also sensed the unfairness of tying their opportunity to take physical education or art or science to performance on mandatory tests—they were speaking up for their right to a public education.

This was the kind of teachable moment I had been waiting for, so I encouraged them to turn this into a story. For the first time since the graffiti story, the boys were very eager and excited. They wrote interview questions and came up with a list of people to interview: fellow students, teachers, the school principal, and possibly the mayor (through e-mail). Their questions were to the point, if perhaps too neutral and objective:

- Why did the superintendent make the Blueprint?
- What effect will the Blueprint have on PE and science classes at Washington?
- What is your own opinion of the Blueprint?
- What will the Blueprint mean for summer plans for teachers and students?

The first blow to the boys' efforts was that the teacher most active in the teachers union withdrew her interview in the face of pressure from the school administration to support the Blueprint. One after another, other teachers also declined to be interviewed. As the immediate excitement of the Blueprint as a headline story faded, students also became less vocal about it, and also avoided being interviewed. The school principal came to our class for an interview, and gave a series of lengthy reassurances to the boys, in words they might not have understood, about how the Blueprint would not affect them negatively and really was a measure to be welcomed since it was for their own good. I watched the boys' faces during the interview: all four of them had the same disengaged, blank expression as they listened to the principal explaining how the Blueprint was a "win/win situation for us all."

Class procedure was to make decisions about the newspapers' layout collectively. When discussing which stories should be on the front page, the boys not only objected to having the Blueprint as the lead story, they declined to run the story at all. "That's not news," they said. "That's all over. It doesn't affect us anymore. The Blueprint is boring." The next week, one of the reporters came in with a poem dedicated to teacher's aides. In the end, it was the only piece published about the Blueprint. The boy who wrote it seemed to feel too insecure in his knowledge of the Blueprint and too confused by the results of his classmates' attempt to make a public statement—although the same writer had done two editorials about topics such as skateboarding. Instead of pursuing the role of reporter, he shifted into the more familiar and childlike vein—expressing gratitude for the individuals he knew would soon be fired:

> Teacher's Aide Poem
> Aides are really cool,
> They can be a useful tool,
> Aides are really tight,
> And when you need them,
> They'll be in sight.
> When there is trouble
> They'll stop the fight,
> And don't worry,
> They don't bite!

In his study of children and the news *The Making of Citizens*, David Buckingham argues that

> Young people's alienation from the domain of politics should not be interpreted merely as a form of apathy or ignorance. On the contrary, I would see it as a result of their positive exclusion from that domain—in effect, as a response to *disenfranchisement*. This reflects the fact that, by and large, young people are not defined in our society as political subjects let alone as political agents. Even in areas of social life that affect and concern them to a much

greater extent than adults—most notably education—political debate is conducted almost entirely "over their heads." (2000, 218–19)

Disenfranchisement is precisely what seems to have taken place when the boys tried to process the Blueprint for the Schools. It is notable that the children's (accurate) definition of the news as very recent, transient events caused them to reject the story that they had felt so strongly about. The students' inability to see the long-term consequences of the Blueprint ("that doesn't affect us anymore"), is also entirely in keeping with the conventions of news reporting. The rap was a brief moment of outrage that the students were able to communicate only in an expressive form.

The most daunting issue for educators is the "comprehensive alienation" of the children from politics, as Buckingham puts it: "It is important to avoid a view of political education as simply a matter of making good the apparent deficits in young people's political knowledge. The more difficult challenge for teachers, as for news journalists, is to find ways of establishing the *relevance* of politics and of connecting the micro-politics of personal experience with the macro-politics of the public sphere" (2000, 221; emphasis in original).

The Blueprint for the Schools incited the children's moral outrage—over the firing of the teacher's aides and the ways poor communities are punished by pro-business policies. But the children abandoned faith in the prospect of speaking up for themselves. Their burgeoning citizenship participation was squelched because it was too difficult to find information to back up their case and the Internet provided little to supplement the blatantly biased coverage of the local newspaper.

Searching for Information

Educational researchers have just begun to study the impact of Internet usage in schools. One of the best studies is Schofield and Davidson's five-year research project on the introduction of the Internet into elementary, middle school, and high school classrooms in a large urban school district:

educational benefits do not flow automatically from Internet access. Attitudes and expectations; technical knowledge; classroom culture and Internet culture; and curriculum design, implementation and follow-through all affect what teachers and students can accomplish with the Internet. In at least one critical respect, the Internet turns out to be no different from any other classroom resource. What you get out of it depends a great deal on what you put into it. (Schofield and Davidson 2002, xi)

The Internet has the potential to empower students by making it possible for them to delve into topics of interest that are not normally covered by school resources. The results of preliminary research on the Internet has noted the possibility of using the Internet to engage students in new, more powerful roles in the classroom that allow them greater independence and more determination over their own education. (Certainly this is one of the goals of the newspaper class, in which instead of assigning topics we work with the children—sometimes for weeks—until they can arrive at a writing topic in which they feel they have a personal stake.) For routine schoolwork, students in the Schofield and Davidson study reported improvements to their work: the Internet enabled them to narrow topics more readily, and find better information. The ability to combine search words was felt to be a significant advance over the school library's card catalog. But the downside of this was that students easily wandered off topic to entertainment sites, and that in the context of a busy computer lab it is impossible for teachers to monitor Web browsing. In addition:

> Information overload may be the biggest problem for younger students. Unlike library resources, the material found on the web has not been prepared with students' level of background knowledge and reading ability in mind. . . . Even students with advanced reading skills sometimes found it difficult to sift through the masses of information they acquired about a topic of interest to select accurate and pertinent information. This in turn meant that teachers often needed to work very closely with students as they used the Internet, particularly at the earlier grade levels. (Schofield and Davidson 2002, 202)

As the experience of the ballpark article shows, critical thinking skills are essential to analyze the editorial ambitions of Web sponsors, the selection process behind the information, and the possibility that divergent interests and opinions have been neglected or ignored. One teacher in Schofield and Davidson's study commented that although she knew that students needed specific training in search strategies and how to evaluate the information students gathered on the Web, she nevertheless resented the necessity of leaving the subject matter at hand to teach about Internet skills specifically.

The success we have had in the newspaper project owes as much to the investment of time and energy by tutors, to one-on-one writing and photography instruction, and to the building of a classroom atmosphere of encouragement and fun as it does to the machines at the students' fingertips. The continuing struggle in my class is how to lead children to a political understanding, how to foster an interest in current events that can be tied to a critical consciousness. Yet these efforts are hampered by the fact that the Web, rather than the fabulous information superhighway to places near and far it was touted to be, offers lots more and easier access to information about the Worldwide Wrestling Entertainment than about school budgets. For all but the highly trained and goal-directed researcher—and the person with access to "premium" subscription content—the Internet is more like a mall than a library; it resembles a gigantic public relations collection more than it does an archive of scholarship. For children whose first language is Spanish and who struggle with spelling, Internet searching can lead to tremendous frustration, and long wild goose chases, unless skilled adult helpers are close at hand.

Concern about the Internet as a faulty educational tool should not be restricted to elementary-school children. Digital network technology has not widened the availability of good information for most adults, either. Some of the blame belongs to news agencies, which are abandoning the public service mission of journalism in the quest for expanding profit margins. Educators must recognize that online computers pose extraordinary pedagogical challenges. Teachers must place the critical evaluation of information at the center of their teaching about the Internet.

chapter 3

Gender and Computer Affinity

Typing versus Gaming

If girls cannot "play like the boys" will this demean their future opportunities in life? Does the prescriptive world view of femininity and the traditional game-playing paradigm, which is most commonly reflected in children's computer games, limit girls' future opportunities?
— ANGELA THOMAS AND VALERIE WALKERDINE (2002, 8)

Social class, ethnicity, and language interact with gender expectations in determining who likes to use computers. Computer affinity develops out of one's school experience, access to computers, and social networks of friends and kin. In order to analyze feelings of affinity toward computers, I observed boys and girls over a three-year period in the Washington computer lab, and over a six-month period at the Clearview lab, and collected field notes written by nearly one hundred senior university students on any perceived gender differences. The university students were observing and assisting at the Washington after-school program as part of their work for a seminar entitled "Digital Pedagogy" (see the appendix). The seminars took place from 2000 to 2003, at a public research university in the same greater metropolitan area as the elementary schools I studied. In addition to tutoring the elementary-school students, the seminar participants wrote essays reflecting on their own relationship to computers: describing their first encounters with computers at home or at school, how members of their immediate family felt about computers, and what influence gender roles played in how fathers and mothers, brothers and sisters embraced, avoided, or outright rejected computers. Because computer affinity is multifaceted, emerges over time, and involves both formal and informal learning, it requires a method that is longitudinal and qualitative and involves the subjects in some degree of reflection on their own experiences (Seiter 1999).

A continuum of computer competence was clearly discernible in the elementary classes, even among students as young as eight, with boys as well as girls falling on the expert and the novice ends of the spectrum. Competence in the computer lab was defined as the students' ability to manipulate the keyboard, the mouse, Web browsers, graphics programs, and word processing programs. While many earlier studies involved classes in which children learned how to program (Kafai 1998; Huber and Schofield 1998; Schofield 1995), my study involves only ordinary, nonscientific uses of computers and the Internet for school and work. Even in this user-friendlier realm, social determinants provide significant incentives and disincentives for the engagement of boys and girls with computers. Boys hold advantages over girls through the association of computers with gaming; but, particularly in working-class and Latino communities, girls hold an advantage over boys because of their easier adoption of the role of good student and their acceptance of typing as a necessary skill for computer use. Whether girls or boys performed better on computers is unimportant—my sample is too small to offer generalizable results, although it does suggest equal competence—the unequal social distribution of computer competence points to the complex ways that age, social class, ethnicity, and language interact with gender identity and the leisure use of computers.

Gender

The spread of personal computers and Internet software since the 1980s has stirred a concern about a "digital gender gap." Lesser interest in computers among women compared to men has been variously attributed to technophobia, to hostile learning environments, to unappealing software, and to a form of gender essentialism in which girls are presumed to be innately deficient with new technologies when compared to boys. Since the late 1970s, researchers have noted sex-related differences between boys' and girls' use of computers and worried about the persistence of girls' feelings of intimidation. In a study of kindergartners in Denmark, play researcher Carsten Jessen noted that "it is generally experienced that the boys happily and energetically plunge into playing with the computer while the girls are more reserved about it. This difference is already clear at the pre-school age, when the gender roles play an important part in the 'computer culture' the children develop" (1996). Jessen notes that the boys and girls develop separate play cultures as young as three. The boys' computer and game culture is informal, extends outside of school, and is very strong compared to the girls'.

In a well-planned study of middle-school students by K. A. Krendl, M. C. Broihier, and C. Fleetwood researchers followed students for three years, controlling for levels of computer experience in making sex-based comparisons. "We had thought that initial responses to computers were influenced by differential socialization and that the effects of sex would diminish over time as children became more experienced with the technology, novelty effects wore off, and computer use expanded. Instead results showed that differences on measures of confidence and interest are quite resistant to such factors" (1989, 91). Consensus in the field has been that girls rarely display the drive and persistence to master computer technology at the same level of intensity as boys do. This consensus framed girls' lack of interest in computing as a problem, a failure prevalent among all but a few exceptional girls. Cynthia Cockburn explains the broader power issues at stake in computer reticence:

> It seems to me that the construction of gender differences and hierarchy is created at work as well as at home—and that the effect on women (less physical and technical capability, lack of confidence, lower pay) may well cast a shadow on the sex-relations of domestic life. . . . Identifying the gendered character of technology enables us to overcome our feelings of inferiority about technical matters and realize that this disqualification is the result not of our own inadequacy nor of chance but of power-play. (193–94)

Like Cockburn, Cheris Kramarae and Liesbet van Zoonen suggested early on that researchers focus on teaching methods, the cultures of computing, the technology itself, or the gendering of computer software, rather than women and girls as deficient computer users (Kramarae 1988; van Zoonen 1992).

As noted earlier, many of the studies in the 1980s and 1990s involved students learning programming (Huber and Schofield 1998; Schofield 1995). These kinds of skills are less important for the general user, and many software programs have been redesigned on the Macintosh/Windows model to make programming operations transparent to the user. Yet research suggests that even after extended work experience with computers, women report less affinity for the machine itself than do their male peers. Despite the fact that for nearly a decade women have held about half of the available jobs in accounting, personnel, labor relations, and financial management—fields heavily reliant on computers—feelings of inadequacy persist among female computer users. In a study of computing activities in the adult workforce, researchers found that even when job level was held constant, men had more positive attitudes about their work with computers (Harrison et al. 2001). The largest difference was on the measure of computer self-efficacy: regardless of expertise and experience, men felt more accomplished on computers than women did. "Males also demonstrated significantly less fear (computer anxiety) and significantly more positive anticipation about computer use. Males found computers to be significantly less controlling, and possessed significantly higher computer self-efficacy than females" (857). One exception to this is that in the clerical job context, where more women were employed than men and the work environment reflected the norms of the dominant gender groups, "females experience more successful computer outcomes than males" (864). This exception seems to verify the extent to which culture and social interaction around computers play significant roles in computer outcomes.

To explain the low enrollments of women in computer science courses, Jane Margolis and Allan Fisher (2002) studied students in high school and university computer science courses. For the university students, they found that "while men do have more experience, prior computing experience level turns out not to be a predictor of eventual success in the program," although it has a significant effect on confidence and comfort. An important nuance of their findings is that even when male students are not especially competent with computers, they do not perceive themselves as lacking in ability. In contrast, women not only perceive male students as knowing more about computer science, but many experience men as doing it with greater ease and more "naturally" (80).

Will this pattern hold true with younger generations more accustomed to computers in their everyday lives and the more pervasive leisure uses of the Internet? Probably not. At the affluent suburban Clearview, these traditional gender behaviors were much more evident than at Washington. Indeed, at Clearview, the boys consistently exhibited more competence and confidence using computers, dominated the class discussions, volunteered for lengthy demonstrations before the class to display their knowledge to others, and created a highly competitive atmosphere. Such observations have been made in studies by other researchers (Schofield 1995; Huber and Schofield 1998; Cuban 2001): boys tend to outnumber girls in computer courses and clubs, and express a very vocal enthusiasm for the technology. The Clearview girls were much quieter, more passive, and more detached from computing activities than the girls at Washington throughout the six-month period I taught there. Most of the Clearview girls were Anglo or Asian; the Washington girls were primarily Latina and African American.

At first glance, the boys at Washington were more territorial about the computers than the girls, and appeared to be more absorbed in what was happening on the screen—at least during the free play sessions. Over time, the Washington lab yielded a more complex picture of gendered uses of computers and other ways of observing children's interactions with computers. Looking beyond the boys' territorialism to actual skill levels, gender differences began to dissolve. As one university student involved in tutoring and observing the class put it (I will refer to such students as tutors in this chapter), the seminar readings had me believing that there would be a significant and noticeable difference in the two genders' attitudes towards using computers, as well as a large difference in the computer usage. . . . [W]hat I found was a more equal usage of computers between guys and girls, as well as a more positive attitude of females when using computers. . . . Both boys and girls seemed to be happy using computers and I didn't see any signs of girls who were reluctant to use computers.

Other tutors identified girls as the top computer experts in the Washington class. One male student who had immigrated from Eritrea and was working his way through college, commented, "I see the girls right along with the boys on the computer every step of the way. In that class I see the strongest person on a computer is Ginger, she is going one hundred miles per hour on her computer terminal, switching from the Internet to her newspaper article back to the net and on to her email account, with ease." Noting that there were novices among male and female students, one tutor observed, "At the university, it is interesting to walk into a computer science course and see a vast majority of men, or walk into a literature or human development class and see a largely female class. Clearly there is a strong slant of men towards computers, yet in our seminar, as well as Washington, gender proclivities were not pronounced."

The equal distribution of computer confidence in my classes is due in part to explicit pedagogical strategies. It is also important to note, however, that the discussion of gender and computing contains a hidden class and ethnic bias. In the United States, the gendered advantage in terms of computing belongs to middle-class males. Anglo boys and men for whom English is their first language. As with other elements of the public school curriculum, the inherent advantage for boys disappears when one considers boys from working-class families.

While the tutors, as students at a prestigious university, represent an educationally advantaged group, many of them were the first members of their families to attend college and wrote reflective essays on their parents' struggles with computers. More complex pictures of gender and computing emerge from their stories, where fathers were fearful or avoidant in some cases. A tutor from a working-class background described her parents this way:

My parents live very simple lives and do not own very many expensive items. I am the oldest of five children and none of my brothers and sisters knows how to use a computer very well, and the little knowledge that they do have comes from the education they are given at public schools. The main reason my parents have not bought a computer is because their jobs do not require them to use one. My mother is a waitress and my father is in construction. They grew up in a generation where there were very few computers and they have absolutely no knowledge of how to even turn on a computer. My father coaches my brothers' sports teams, and he always brings the team roster and announcements to me to type. Together my parents believe that if they own a computer then the kids will become addicted to it and it will cost them more money to keep it upgraded. My parents are also hesitant to buy a computer because of their limited knowledge of what type to buy, how much they should spend or how to set it up.

In another tutor's family, the mother's job at a high school was the driving force behind the introduction of computers into the home. (Nationally, teachers were among the earliest female adopters of e-mail and Web browsers, and accounted for the first surge in women closing the "digital gender gap.") The tutor explains:

> My father is the oldest member of our family, and he has the least experience with computers. I believe that there is a connection between the two statements. As is the case with many older people, he didn't start using computers until later in his life. My father grew up on a farm for his early childhood years. Obviously computers weren't in the equation, and this kind of lifestyle didn't tend towards "keeping up with the Joneses." He moved to San Francisco when he was a bit older and attended UC Berkeley but computers didn't play much of a role in his life at this time. . . . So how did he become more techno-savvy? There were a number of factors, especially my mother. My mom is currently a high school librarian in San Francisco. This played a big factor in the inclination of my family towards computers. I remember that during long breaks, such as summer and Christmas, she would bring home Apple computers. My brother and I would fool around with the educational software, and she would see how good the program would be for students of our age.

Other tutors recounted how adapting the keyboard for Chinese characters proved daunting for their parents or grandparents, who refused to learn e-mail or Web browsing. Learning to use the Web or e-mail is made enormously more difficult for parents or grandparents for whom English is a second language and whose first language uses Chinese characters or other writing systems such as Arabic or Cyrillic. This was reported by many Asian American university students, the group most often stigmatized in California as having an unfair advantage in admission to public universities. Consider this story by one tutor about her father, who was a professional, and worked in a technical field, but was still daunted by language barriers and embarrassed to appear deficient as the father of the family:

> My father has a Ph.D. in biochemistry. He is very science- and academia-oriented, and not a very technical person at all. Even when my brother would buy small robot construction kits, my father would be reluctant to try to answer his questions, although he would try to show it up by pretending he was "too busy." His work requires very little computer usage, and although many scientists today are heavy users of computers and the Internet, my father has managed to get by without doing anything he feels intimidated by. He is also not a native speaker of English, and although his understanding of it is very good, he is still more comfortable with his native Korean. His peers are all fairly comfortable with computers, and he has friends who work in the computer business, either as importers, retailers, or software programmers and the like. Even so, he is quite reluctant to really sit down and try to figure everything out, as I believe it makes him feel incompetent, and his social class and high level of education might contribute to his feeling that he must appear competent in all things at all times.

These same feelings of inadequacy were reported in Liz Stanley's interviews with nonusers in San Diego County (2001). Often, the desire to avoid the embarrassment of appearing incompetent in front of other family members was a crucial influence on fathers of Mexican descent, who tended to avoid touching computers altogether.

At Washington, the boys' greater enthusiasm for computers did not translate into focused work on the computer or leave the impression on their observers that they were especially gifted in the realm of technology. Tutors' field notes consistently remarked with surprise or irritation how challenging they found it to keep the boys on task, even with a student-to-teacher ratio of three to one or two to one. Two tutors reflected on the girls' greater seriousness about completing assignments:

> It was interesting to note that the girls seemed to be much more focused in finding their information, and they were much more business-like in their use of computers. Guys would often become sidetracked and go to websites that weren't necessarily related to their topic. While some might argue that these boys were more into computers and the Internet, as evidenced by their enthusiasm for web surfing, I believe that girls were simply more focused on their tasks.

> At Washington the girls did use the computers as much, but the boys dominated the usage of them. The girls were more interested in writing their compositions whereas the boys seemed more occupied in using the computer but not necessarily for academic means.

For many of the boys at Washington, learning the keyboard was an activity that seemed boring compared to Web-browsing, game playing, or even layout or digital photo manipulation. One tutor describes her frustration with an African American fourth-grader when she had to prompt him to finish his news article:

> When I did get him to type his article, he was easily distracted. Rather than working on his article, he would open the Kid Pix program and start to draw. Other times, instead of him typing, he wanted me to type. He felt that typing took too long and that it was boring. I noticed that he knew the location of all the letters on the keyboard, yet he only used two fingers to type. Typing was a struggle for him and he easily became frustrated.

In fact, many of the middle-class Anglo women who participated in the university seminar found this trait in the elementary school boys—most of them African American and Latino—to be vexing. Boys of color do not fit the hacker image, and their resistance to doing assignments was never interpreted as a sign of cleverness. The women tutors found the boys' refusal to type annoying and sexist: for the Washington boys, whose fathers work as mechanics, plumbers, construction workers, and security guards, the sight of men typing is rare. For the children at Clearview, whose parents are white-collar professionals, it is much more common to see adult men typing on a computer, and keyboarding skills are part of the regular curriculum starting in the early grades (although even there, teachers report

Illustration 3. Boys often demanded help with typing.

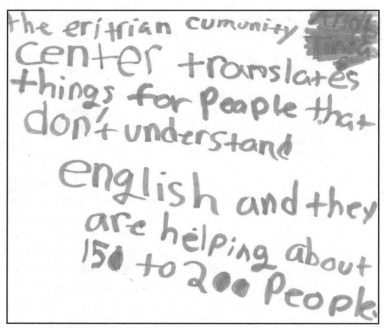

Illustration 4. Typing was a greater struggle when English was a second language: these are notes for a story about the Eritrean Center.

a greater resistance among the boys to learning touch typing). The Washington boys clearly assigned typing to a realm of feminized activity: this attitude was common among adult men in the area as well (Stanley 2001). When I would urge the boys to spend time on the typing games such as Mavis Beacon Teaches Typing, they openly resisted, explaining that they did not need to learn typing. Frequently, boys would try to cajole me into typing their school assignments for them. The boys surmised that teachers give extra credit for a typed assignment, but flatly refused to take the time to improve their skills. In this respect, the boys recognized the historical and cultural association of typing with feminine labor, even if it is happening on a computer. Cockburn has called this "the intrinsic interdependency of keyboard and computing," wondering if it will eventually force a "re-gendering of typing so that it is no longer portrayed as female" (1999, 195). In the white-collar field of university teaching, this re-gendering has already occurred to some extent, so that the association of typing with female labor had become nearly invisible to me before the Washington students displayed such a keen awareness of it. Feminist historians of technology, however, have been careful to trace the complex ways that typing was integrated into computer software. Jeanette Hoffman has studied the gendering of early word processing systems:

> software development—which, at least in principle, is free of physical constraints inherent to its mechanical predecessors—shows a clear tendency to reproduce traditional socio-technical arrangements . . . whereas dedicated word processing systems treated female users as eternal beginners, programs like WordStar and WordPerfect epitomized the opposite view according to which secretaries could be expected to learn even complicated interaction languages. Amazingly enough, the least competence regarding all kinds of application software was attributed to men. . . . The meaning of gender—portrayed as imagined skills, competencies, and tasks—varies depending on concrete practical circumstances of software development and use. (1999, 240–41)

One African American student in my seminar, whom I will call Monica Hughes, related a poignant story about her greater academic success compared to that of her younger brother. Monica attributed much of her academic success to the fact that she was enrolled in the James B. Flood Science and Technology School. There, Monica enjoyed the latest technology, daily computer classes, and abundant free time on the computers. "Our typing program was made into a game of competition, which motivated us to learn to type. If you made a mistake, the sprinter would trip over a hurdle racing towards the finish line." She compares the high motivational value of her education to that of her younger brother, James, who attended the regular schools in the Ravenswood district, in East Palo Alto, known, as Monica described it, "for its lack of funds, low test scores, and students graduating unprepared for high school in a community infested with drugs, violence, and gangs."

There, James's only exposure to computers was the old Apple II her mother could provide. Her brother's grades and self-confidence dropped in high school.

> When given the assignment to do a simple book report, reading was not his issue, but rather typing out the report, a requirement of the assignment. James says he just "became frustrated and didn't want to do it." James would rather not turn in a report than deal with the ordeals of typing it out. Doing research reports turned into a process where my mom would have to "hold his hand" throughout, a task difficult for a busy mother of five to take on. Joshua's confidence in school dropped due to his lack of computer literacy and many assignments were left undone.

Thus working-class boys suffer a double disadvantage in the contemporary environment where teachers expect typewritten assignments starting in elementary school—both because they do not have the latest computers with word processing at home and because they have few male role models who know the keyboard.

Certainly many male undergraduates recognize that their computer efficiency would be greatly enhanced if they had learned touch typing. In writing about their families, several undergraduate students noted their own fathers' recalcitrance in matters of typing, and roughly half of the male undergraduates used "hunt and peck" methods, while nearly all of the female undergraduates had picked up touch typing in elementary school or high school. Typing was a struggle for the girls at Washington, too, but their desire to be good students kept them trying more persistently than the boys, as indicated in one of the tutor's field notes:

> Sherall was typing out her rough draft rather than looking up web sites like the boys. She was trying to complete a final draft and it showed on her face. Particularly, she seemed very nervous and extremely careful in typing out her words. It seemed as though making a typing mistake was the last thing she wanted to do and she used her facial expressions and grunts to kind of prevent that from happening.

For Latino students, typing seemed to add considerable pressure to the academic task. Unlike writing on paper in one's own handwriting, committing one's work to a computer screen meant that mistakes would be visible for anyone to see. In this environment, where a student will frequently find another looking over his shoulder while he is typing, the pressure is increased, especially for non-native speakers of English. One observer noted:

> Many of the boys seem to shy off when it is time to use the word processor. Once they are on the computer and typing there are fundamental basics they often don't quite grasp, such as using the return key or using the shift to get capital letters. These little things can make them feel stuck and behind. It also seems that since a lot of them are second language learners, they may be afraid to show their English on the big computer screen.

This tutor, a woman who was studying for a teaching credential and exhibited more empathy than students in the class who had little experience with bilingual students, wisely recommended that all emphasize continually that the first draft does not need to be perfect:

> remind the kids that the first draft doesn't have to be perfect. In fact it is just an exercise to get your thoughts on the screen. Giving kids reminders such as this will hopefully take away the fear that they will look stupid, because if everyone is making it rough then they are just another in the crowd who needs to edit their paper a few times. . . . The language barriers will also be worn thin by encouraging mistakes on the first draft; once again it is not necessary to worry about mistakes if everyone understands this is part of the task.

This was especially true in the Washington class, where the appearance of type on a computer screen was sufficiently novel to be greeted as a kind of spectacle by the students, who would often stand around just to watch the words appearing on the screen. It is a testament to the relative novelty of typing in their everyday lives that boys and girls alike never tired of watching me type. Crowds of six or seven students would gather around any time I was seated at a computer typing rapidly. This tangible skill impressed them more than my occupation, or my educational credentials, or the car that I drove (another subject of considerable curiosity), or the books I had published. They perceived the speed of my typing as miraculous, but they also enjoyed watching the screen to catch any errors I had made in spelling or punctuation. At Clearview, where most parents hold white-collar jobs, my typing attracted the students' attention on only a single occasion.

Now that chat rooms, e-mail, and instant messaging are so popular among children, learning to type offers greater rewards, in the form of interaction with friends. Despite the abbreviated style of most instant messaging, the activity requires mastery of the demanding and distinctive form of typing required for this fairly unique phenomenon of "rapid written communication" (Guy Merchant quoted in Willett 2003, 2). Rebekah Willett's research on chat rooms and language play used by immigrant children at an after-school program in London suggests that learning is swift and occurs "just by interacting on the sites and reflecting on the sound of their spoken voices in an urgent attempt to enter into the fast pace of chatrooms" (Willett 2003, 12). But these forms will be much harder for those who are uncertain of spelling or language usage. Instant messaging is also primarily available to those who have private computer and domestic Internet access, so there is also a substantial economic disincentive to children in working-class families. It is more important than ever, however, for working-class children of both genders to master reading and writing, because the Internet, for all its flashy graphics, is primarily a linguistic medium. Forget about computer literacy: literacy is more important now than ever.

Gaming

The relationship between boys' comparatively higher interest in computer games and their comparatively larger representation in high-power computer jobs is not accidental. Computer and video games provide an easy lead-in to computer literacy . . . and so those children who aren't playing them at young ages may end up disadvantaged in later years.
—JUSTINE CASSELLS AND HENRY JENKINS (1998, 11)

If typing can function as a disincentive to using computers, the association with gaming pushes toward a durable gendering of computer technology as masculine among children. The proclivity to play with computers is tightly connected with the leisure sphere of video gaming, while task-oriented computing, such as word processing and spreadsheet programs, are closely associated with the labor sphere of clerical work.

An interest in gaming paired with a lack of domestic Internet access was responsible for bringing many of the male students to the after-school computer lab. When we distributed quarterly questionnaires to the students asking about their favorite features of the after-school class, boys most commonly answered playing games, while girls' responses often listed the experience of being a reporter—doing interviews, getting one's name in the paper—as the best part of the class. Allowing free time on the computers to pursue this interest kept the boys coming to class: when occasional substitute teachers imposed stricter work standards, the boys' attendance typically dropped off until I returned to the classroom. For working-class Black and Latino boys, game consoles are more familiar domestic objects than are personal computers—and gaming is clearly associated with working-class masculinity, if the vocational tracks are filled with middle-class students (Stanley 2001). There were educational costs as well as benefits to the permissive rules regarding gaming in the after-school class.

At times the boys' pursuit of games and game culture was another means—along with refusing to type—of resisting teachers and interfered with their performance as students in the class. Consider the following scene from my classroom. One day during the fall of 2001, I watched as two of my tutors presented a lesson in the Washington computer lab on how to import Web graphics into Kid Pix and then manipulate them to make your own picture. The undergraduate students were nervous, and unprepared to translate their computer knowledge into words the elementary-school children would understand—they assumed that terms such as *URL, cursor, search engine, application,* and *graphics* would be part of the children's vocabulary. Most of the children did know these concepts, or could learn them in the space of minutes; however, they did not know them by the same terms the undergraduates used—or even by any terms at all. As normally happens during technical demonstrations, there was an eruption of Spanish speaking as the children switched from English—the mandated classroom language—to Spanish when offering help to one another behind the computers.

The undergraduates guided the students to a free clip art site called Ditto.com, and asked the students to search under "American Flag." The most mature boy asked if they must stick with the flag, or if different images could be chosen. The undergraduates hesitated, wondering what kind of substitution he had in mind. The boy told them he would rather do Pokemon—and immediately seven more boys working on computers on that hub switched to Pokemon or Digimon characters. They were all fifth-grade boys, three Latino, three African American, and one Somalian. (As always, the children are lined up at the computers according to strict gender segregation.) For these boys other icons—the flag of Mexico, the wrestling star the Rock, the rapper Nelly—have more meaning than the U.S. flag. As the boys continued, they refused assistance from the "floating" tutors, who were mostly women, until one savvy tutor tempted the boys with an offer to show them how to make Pokemon characters look meaner, uglier, or nastier. At this point, all of the boys gladly accepted her demonstration of the features of the software.

This story exemplifies many of the trends that I observed in the computer lab each class period: boys seem compelled to use the World Wide Web to indulge their fan interests, and they are gratified with what they find: an endlessly proliferating, varied collection of images and information related to their favorite video games. In this regard computing and gaming go hand in glove. During group exercises such as this one, boys are more independent than the girls, and more likely to be doing something other than the assigned task. In the collage exercise, the boys resisted the instructions of the teacher, and also resisted the help of the coaches, preferring to work secretly. They were flatly uninterested in the assigned example, with its post-9/11 patriotic flavor. Girls rarely wandered so far from the instructions they were given during a class project.

For boys, however, computers are so strongly associated in their minds with games and popular culture characters that they seem unable to resist searching for them each time they use computers, thus reinterpreting class time as leisure time. The boys' actions in class, as reported in the field notes, reflects some typical behaviors of same-sex peers at school, as the late psychologist Beverly I. Fagot noted, in a study of same-sex peers at school:

> Boys were influenced by other boys, but less by girls or by teachers. Boys were not influenced at all by girls or teachers if they were engaged in male-preferred activities (defined here as play with masculine toys or rough and tumble play). In other words, boys engaged in male-preferred activities appear to have developed a group structure that resists the social demands of anyone except other boys. (Fagot, Leinbach and Hagan 1986, 209)

Carsten Jessen has noted that boys "take great pleasure in deliberately doing things wrong." Girls are more rule-abiding and attempt to get things right. The boys' pattern is self-reinforcing because it is more experimental and results in a greater exploration of the software (2000, 11).

Lesley Haddon's observations based on time spent in a computer club similarly suggest that girls, partly because of their weaker interest in gaming, do not use computers as a topic of conversation at school even when they do use computers at home. Instead, girls are much more likely to discuss movies or music with their peers (1992, 91). As one tutor at Washington commented in his notes, "I feel computers are a masculine activity that is difficult for both males and females. However, I also feel that males are more into computers for two reasons: media preference and physical location. . . . Boys would rather watch television and play video games than read and listen to music like girls."

The girls in Haddon's study were also unlikely to play games on computers in public places, stores, arcades, or school clubs, but used them at home. In practice, it was indeed male electronic and computer hobbyists who took the lead in developing the microcomputer, and they did produce the early systems mainly for other male hobbyists (1988, 253). Haddon points out that the significant factor about video games is that many males learned to play them in arcades, where watching others play, learning tips, and talking about the games was a part of the culture. When video games migrated to the home computer, this culture went with it. While the stereotype of the computer hacker is male and a "lone wolf" type, in fact computer enthusiasts rely on a great deal of peer tutoring. Marsha Kinder (1991) reports that 90 percent of the readership for video game magazines are boys—and it is this kind of extra commitment to reading about the machines that is necessary to master more complex operations on computers. In contrast, girls were not welcome in the arcade culture and had a very different set of social behaviors around games, even when they played them at home.

> It is a myth that girls do not play games. It may be true that fewer pay games and do so less intensively than boys. But the distinctive feature is that it is solely a domestic experience for girls, who do not discuss it so much outside the home or read about it so much in magazines. . . . [B]oys who take an interest in games have an additional culture outside the home that supports and reinforces their interest. Hence they are more likely to appropriate the machine, to create demand for computers within the home, and to influence the choice of games software entering the home. (Haddon 1992, 254)

When boys are avid gamers, they also demand the most sophisticated hardware available. Indeed, computer gaming is one of the few market forces driving the acquisition of new computer hardware, now that most non-game software comfortably runs on existing processing power. One Chinese American woman tutor—who was highly competent on computers but did not enjoy using them—reflects on gaming access and exclusion in her family:

> Fred, my brother, is the all around superstar angel of the family, not to mention he is the first-born son. Being the oldest, he was given a brand new computer with a color monitor. Ditching his old joystick Atari set, Fred moved on to computer games. He has no apparent anxieties about computers and used them with ease and confidence. While Fred was given the new computer, I was stuck with the ancient green-and-black screen Apple. The graphics were limited and the games were not very challenging. I often ended up on my brother's bedroom floor watching him play challenging, problem solving, combat games. These computer games held much appeal even though war, combat, domination, control, and competition were central themes to the games. I was given little opportunity in playing these kinds of computer games. I was limited to watching someone else play, or solitaire. Eventually I lost interest in computers and turned to reading.

Nearly all of the girls at the Washington lab enjoyed playing some of the online and CD-ROM games. As newcomers to the lab, most girls went through a period of intense absorption with Bugdom, which came factory-installed on the iMac computers. The game involves guiding insects through a mazelike environment filled with naturalistic obstacles such as tall grass and predatory creatures in a vivid, three-dimensional format. Bugdom employs some of the features that game designers have suggested appeal to girls: it is nonviolent and does not have dramatic negative feedback, such as character deaths. There is a musical background, but the soundtrack does not accentuate the pace and create arousal (Kafai 1998). Bugdom was a favorite among many girls, especially "lone girls" unconnected to one of the groups of "popular" girls. Girls throughout the school year also regularly picked up Myst, although in the absence of any expert players no one advanced beyond the lowest level of the puzzle, and thus the girls were frequently caught in a loop of repeating locales.

Susan, a ten-year-old immigrant from Honduras, was amazed that she would be allowed to play computer games in school, and during her first weeks in the class often pleaded to play instead of writing her articles. Eden, from Eritrea, was also drawn to Bugdom, and preferred to play rather than struggle through writing and spelling. But Susan and Eden lost interest in Bugdom after two months, whereas many boys played the game intermittently with great enthusiasm for periods of two years or more. This is typical of the scaffolding problem with video and computer games: novices without more knowledgeable playmates remain stuck at the lowest level of the game, never accessing more interesting levels, worlds, and challenges, and therefore utilizing less than 20 percent of the games' capacity. Scaffolding is what lends video gaming its air of intense competition and its elaborate rites of secrecy and initiation: players must commit hours to practice playing the games if they are ever to advance to higher levels.

As an experiment at the Washington lab, a summer session was organized in which students had free lab time, with no obligation to work on the newspaper. Children could choose their own activities, and a selection of about twenty CD-ROM games were offered. Boys found the availability of the games and the unlimited time to play on the computers to be a huge draw. Girls enjoyed surfing the Web and playing the online role-playing game Neopets (discussed further in Chapter 5). They enjoyed drawing and talking to the teachers, however, as much as the games. Some girls never played a game after the first day of the summer session. When girls did play the games, they tended to stick to the "pink" software. A nine-year-old Latina girl named Alexis was absorbed in a single game, Barbie Fashion Stylist, and spent hours painstakingly re-creating the precise costume, hair, and makeup of the dolls on the cover and instruction booklet of the CD-ROM. She never innovated styles, but set for herself the goal of exact mimicry of the packaging. None of the other games appealed to her, and she was always eager to drop the game to draw pictures, or chat, or look at websites discovered by other girls in the class that related to celebrities or music.

Ten-year-old Mary, another girl of color, showed considerable ingenuity in finding sections in the All About Me CD-ROM game (a product of the feminist pink software entrepreneurial movement of the 1990s) that appealed to her interests. She had done a story for the paper about the local Wicca store, and was fascinated by superstition and the occult. Her mother was Filipina and her father Mexican. She found a fortune-telling section in All About Me and was typing in questions about her future to the psychic, "Signora," a generic figure of Orientalist exoticism. All About Me was more familiar than the sports games Backyard Baseball or the medieval, fantasy, and anime characters who populate so many video games. Like Daisy, however, Mary's attention to the game was always secondary to her interest in interacting with the teachers, other girls, or the pursuit of her popular-music interests on the Web.

In Australia, Angela Thomas and Valerie Walkerdine organized an after-school computer club to study gender differences. This club focused on game playing, and given the popular wisdom that boys are more avid players than girls the researchers were surprised that girls signed up for the club in equal number to the boys and that they were "enthusiastic and excited" to be part of it. Over time the girls showed that they had similar levels of game knowledge and even skill compared to the boys. Yet gender differences resurfaced when the data were analyzed more closely:

Despite our hopes and expectations that girls who saw themselves as a part of the gaming culture would be able to showcase their keen intellect and skills when game-playing, this was rarely the case. With great disappointment, the observations, fieldnotes and video data all pointed to traditional female stereotypes and a "lacking" of something that the boys did have but the girls did not. . . . Boys would totally immerse themselves in a game, they would engage imaginatively with the characters, their talk around the game would include amusing embellishments on the game action and they would persevere with the technicalities of a game until they had successfully mastered, completed and achieved their goals. (2002, 3)

The Australian study marks a generational change, in which girls are less averse to identifying in some level with gaming, and have attained greater mastery of some gaming techniques. Yet there is a marked gender difference in the intensity and the drive to pursue gaming, which mirrors the field notes gathered at Washington.

Other studies confirm gendered differences in children's computer use and orientation to gaming. Yasmin Kafai organized a project in which children at an inner-city Los Angeles elementary school were taught to program their own games. Clear-cut patterns emerged as to genre preferences that reflect this differential level of interest in video games. Kafai categorizes the work of her students into three types: action/adventure, sports/skills, and teaching context. Girls were much more likely to choose "teaching context" games, with familiar everyday characters and settings; boys were much more likely to choose action/adventure, with fantasy settings and violence. Kafai suggests that one reason girls might have chosen educational games is that they were mimicking the choices of industry producers, who usually market games to girls as a learning tool. "Another interpretation is that girls might have simply followed the directions of designing an education video game down to the letter of the word" (1998, 100). This interpretation of the data seems correct, based on the results of our observations in the lab. Kafai notes that there is more flexibility in gender preferences in play: she warns that "boys and girls find many (but not all) of the same game features appealing" (93). The most striking tendency among the girls was occupying the position of the good student, which was the choice made by many of the Washington girls as well, and none of the Washington boys.

Girls never suggested video gaming as a topic for one of our newspaper articles. The boys consistently chose to write about video game technologies, and they were always able to form groups of three or four boys to coauthor the articles. At Washington, few boys had the money to buy the latest consoles and games, which were getting substantially more expensive during this period, with the introduction of PlayStation 2, Game Cube, and X-Box. Most of the children had old Sega Genesis systems at home, and a few had Nintendo 64. There was intense enjoyment, however, in using the Web to look up information about gaming, and to examine products that they had never seen. In the district surrounding Washington, no stores sell high-priced electronic goods, there are no Blockbusters, and no large chain stores such as Target or Wal-Mart are nearby. Often, video games were known only through having heard about someone seeing them—that is, until the boys realized how much information was accessible on the Web, from Sony- and Microsoft-hosted sites to eBay and Toys "R" Us.

The boys produced some of their best work as journalists while writing about video games. They easily brainstormed ideas and enthusiastically put their research skills into practice: phoning local stores to check prices and inventory, doing vox pop interviews at the local park, and reading up on the business of video gaming, sales figures, and future trends in the *Wall Street Journal* online (at my suggestion). More importantly, video gaming opened up the most explicit critique of the entertainment industries and the strongest class discourse in relation to the consumer goods usually celebrated in the class, as evidenced by the following article.

> Playstation2: Worth the Money?
>
> Kids want Playstation2 but it costs $375. Some people were selling it for $600 or more at Christmas because the stores were sold out of them. The demand for Playstation makes people cheat other people by selling it for hundreds of dollars more than it costs. Sony has said that the price should drop to around $189 once production is caught up. . . . Why does Playstation2 have to cost so much? One student commented: "Kids want it so bad, they make their parents buy it, even though it costs a lot of money."
>
> A lot of kids say that Playstation2 is better than regular Playstation because you can play DVDs, CDs and computer games on it. Other kids say that there are only a few things different from the original Playstation. . . . The Playstation2 games cost about the same as games for Nintendo 64.
>
> There are many complaints about Sony and Playstation2, besides the frustration at not being able to buy one. One problem is that kids get frustrated, knowing that soon they won't sell the old ones any more. Jacob Reeves wondered, "Why do they keep making the technology more advanced when they just make it more expensive? Why don't they just make it right the first time and forget about making new ones?"

The video game stories were comfortable for the boys—a topic that fit their developing notions of masculinity. It sustained their interest, and it generated lively discussions about planned obsolescence and the profit motives of Nintendo, Sega, and Sony. Because their limited cash had left them with obsolete consoles and retired platforms, there was a righteous sense of outrage in their discussions, of the unfairness of generating desire in kids and pressure on budget-conscious parents. In discussions of video gaming, the children recognized the planned obsolescence that stimulates consumer purchase but, for families unable to keep up with the latest, supposedly greatest technologies, excludes much of the population from participation in up-to-date cultural referents. All of these things were precisely what I had hoped to achieve by allowing kids to research their leisure interests and develop critical perspectives on the media industries. In writing newspaper stories about video gaming, the boys shared their passionate interest with student readers of the newspaper, and were able to smoothly collaborate on the research and writing for the articles. The unintended consequence of video gaming as a topic was that it separated the boys and the girls. Girls never joined one of these writing groups, and were not engaged as experts, even as interview subjects. Furthermore, video games continued to be singled out by teachers and the school principal as the most shameful misuse of family money: they often mentioned to me that they could not comprehend why a family with few books and no computer would purchase a video game for their son as a Christmas or birthday gift. This is parallel to Walkerdine's position that it is only working-class boys who are stigmatized for their gaming, while in fact middle-class boys play more and have more money to buy the games and consoles (Walkerdine 1999, 10).

Of Tech Gods and Improved Pedagogy

Some computer researchers have noted that girls thrive in environments where equal time on computers is ensured through careful monitoring, and that "gender differences between boys and girls disappear when computers are used for a real rather than a fantasy goal and involve interpersonal cooperation" (Huber and Schofield 1998). I developed the newspaper class/computer curriculum with this kind of intervention in mind—not computer use isolated from other endeavors but instead computers used as one of a set of tools among others for a specific purpose. So while the boys in the course typically referred to the program as "computer class," and listed their favorite components of the class as the opportunity to play games on the Internet, the girls typically called it a "newspaper class" and enumerated the roles of reporter, writer, and especially interviewer as their favorite components of the class. I had planned the after-school course as an intervention to get girls interested in acquiring computer and Internet research skills. While the girls also listed Internet access and using computers as positive factors, their primary roles were as reporters and photographers rather than computer whizzes. This boosted girls' enrollment and attendance, and allowed them to clearly establish their own culture in the lab. The atmosphere in the Washington computer lab was lively and sociable, as one tutor described it:

> Computers were designed to be a personalized and solitary activity: one keyboard, one small monitor, and one mouse. In a school lab computers are a very social activity. At Washington, the children teach others how to save, give instructions on where to find pictures of Shakira, or show off to others their creativity. For example, Jacqueline had written an excellent poem, and at least five children had crowded around her computer to read her poem. Or when someone is playing Bugdom there will often be someone else watching the game.

Though the presence of computers was initially an enormous draw for the boys, the girls were most intrigued by the possibilities for downloading fan material and songs and music videos from the Internet, playing their music CDs on the computers, and gleaning information about their favorite pop stars. Because the class emphasized verbal skills such as good interviewing technique and writing, the girls were never at a disadvantage in terms of the overall package of skills that were cultivated and rewarded in the class.

In class enrollments during the first year of the program, boys outnumbered girls by more than two to one, but as time went on enrollment grew through word of mouth among friendship circles of girls. Mothers in the neighborhood often went to great trouble to ensure a place in the class for their daughters, including bringing a friend or sibling to translate discussions with the class administrator, and coming to the lab on foot with several younger siblings in strollers or in their arms (see illustrations 5, 6). Quarterly talent shows became a regular feature of the class—part of the class celebration each time a new version of the newspaper was completed. More so than any other, the talent shows helped to widen the class enrollment, as visitors entered the computer lab initially as audience members for these student performances. While there was risk involved in spending so much time on an activity that was not directly related to literacy or computing, the talent shows built a strong sense of classroom community and clearly enticed new girls to sign up for the class the following term.

Margolis and Fisher found that the most effective strategy for getting girls into computing was recruitment of friendship circles, which helps compensate for some of the recurrent disincentives for girls in computer courses, such as the lack of female students, male students making them feel incompetent, the risk of seeming geeky, and the fact that most computer games bore them (2002, 113). Margolis and Fisher identified one common trait among women who persisted in studying computer science at Carnegie Mellon University and reached skill levels at which they began to perceive themselves as competent: they were able to find alternatives to the dominant image of students as male and as "narrowly focused, intense hackers" (133). In fact, it was the male hacker image that women found most repellent and discouraging in computer science courses. Margolis and Fisher's project was to test specific educational interventions to improve the success rate of women students. When faculty and students were made more conscious of the ways the "boy hacker icon" influenced their course and project design, as well as the culture of the computer lab, a more multidimensional view of computing could be nurtured that was more inviting to female students.

The research project of Margolis and Fisher included a significant outreach component to high school teachers of computer science. In special summer sessions, the teachers were asked to analyze the reasons girls stayed away from the classes and to plan strategies for making the class more widely known as an elective course girls should at least consider taking, and make the course and the learning environment more attractive to girls. Margolis and Fisher warned that for teachers to bring about

Illustration 5. The students usually segregated themselves by gender at the banks of computers.

Illustration 6. Recruiting friendship circles led to a high enrollment of girls in the class.

such a change required "a ferocious attention to the quality of the student experience" (2002, 140).

One dimension of student experience is the lack of public recognition of female expertise. From universities to elementary schools, I have observed that both male and female students confer on male students the status of expert or "tech god" (Cuban 2001) and turn first to males for help. In a study of California high schools, Larry Cuban identified two emerging types of students:

> There were a handful of students whose in-school lives changed with increased access to technology. We called them "open-door" students (their computer competence enhanced their desire to do well in school and hence opened doors to learning) and "tech gods" (students recognized by adults and fellow students for their substantial expertise). Open-door students were predominately, though not exclusively, male and from varied ethnic backgrounds. (2001, 91)

Tech gods, according to Cuban, were exclusively male and usually Anglo or Asian. The elevated role of the tech god in the status system of the school has increased due to the chronic shortage of funding for computer technical support in K–12 schools as well as in universities. Teachers and administrators rely on students to repair, connect, and maintain school networks, paying them with independent study credits and free periods instead of cash (92).

To correct the tendency to identify only those tech gods who are male, I instituted assignments in my college courses in which women are especially encouraged to take over the teaching during hands-on demonstrations and to act as coaches for the other students. In my university courses, students have positively evaluated such requirements, and many women went from initial reticence to taking on a substantial role as roving helpers during our sessions in the computer lab. At the elementary school level, however, girls were much more reluctant to emerge as classroom leaders. Only one girl, over the course of four years, sought out and promoted herself as a technical expert: this was the Anglo, working-class Washington student Ginger, who was obsessed with fantasy games such as Neopets and identified herself as a "computer geek." The girls at Clearview consistently refused to play any leading role in the classroom.

A more typical girl than Ginger at Washington would be Olga. Some of the most competent girls, especially those who were Latina and among the most respectful and considerate of other students, were far more reluctant to identify themselves publicly as knowledgeable about computer operations. Olga was brought to my class the first day by her teacher, who helped recommend talented students and secure parental support for the after-school class. Olga was from Guadalajara, Mexico: she spoke English with a strong accent and never spoke Spanish in class. She was small and perfectly groomed, wearing very modest clothes. She was a model student: quiet, attentive, and always on task. One of her classmates was Jason, an Anglo American boy, one of only three in the class that year. Jason was overweight but wore trendy clothes and flashy hairstyles. He talked whenever the teachers talked, left his seat regularly, never worked unless prompted to or unless threatened with expulsion from the class. During the hands-on exercises at the computer lab, there was an onslaught of demands for assistance, especially among the Anglo male students. Although the Anglo boys were a minority of three in a class enroll-ment that varied between eighteen and twenty-five, they seemed to feel entitled to a great deal of attention from me and the university students.

Olga and Jason frequently ended up in seats back to back at the computers. For months Jason demanded incessant attention from the teachers. Finally, I realized that there was someone there, Olga, who understood Kid Pix better than I did and had offered to help Jason on several occasions. He would not accept her help. Olga was so quiet and modest in class that I had not noticed her proficiency on the com-puter. Instead it was called to my attention by one of the seminar students. Because she rarely spoke up I had the impression that she might be struggling. By the time I finally noticed Olga, among the flurry of activity and demands, she would often be sitting at the computer doing nothing but waiting. Unwittingly, I had stereo-typed Olga as a very obedient and passive student, and this led me to assume she had not yet done the assignment, instead of surmising that she was speeding through them at a pace much faster than her peers. Olga needed greater challenges than the assignments we gave her, or a promotion to the role of roving helper. Meanwhile, Jason let out a constant stream of demands, and became very disrup-tive when we ignored him; he usually got the attention he clearly felt entitled to. But to see this only as a gendered issue would be to simplify the issue of who is like-ly to become a tech god. Because mine was a longitudinal, ethnographic study, I had more information about Jason than could be gleaned from this particular observation alone. He spatially separated himself from the Latino boys in the class, and frequently fought with the African American boys in the class. In doing so, he also isolated himself from several boys who were far more competent than he. On a worksheet asking students to describe their community, Jason defined his as "other white people." So Jason's avoidance of Olga had to do with ethnicity as well as gender.

As teachers, we need to look out for the Olgas in our classrooms. We also need to question which students are viewed as smart about computers and why. There is a tendency to ascribe intelligence to boys who behave in certain ways and dis-play certain skill sets. The boys at Washington were brought to computers from an interest in games, but unlike their Anglo middle-class peers at Clearview, their interest in gaming did not nominate them as future tech gods or hackers—it was just seen as part of their poor performance as students. Neither did their refusal to learn typing serve them well. Let us try to even out the advantage boys enjoy over girls because of the link with gaming culture, but let us also be ready to recognize other kinds of skills—especially those tied to writing—and imagine new ways of making them meaningful for working-class boys of color. Overall, the univocal concern with the gender gap no longer suffices for understanding the breadth and complexity of today's digital divide.

chapter 4

Wrestling with the Web

Latino Fans and Symbolic Violence

For nearly four years, children were heading to my computer lab as soon as school let out to explore their interests on the Internet. Many fads came and went in the years from 2000 to 2003: Pokemon, Digimon, Yu-Gi-Oh!; Neopets; Powerpuff Girls; free games at Bonus.com or Nickelodeon.com; music videos of N'Sync, Backstreet Boys, Aaliyah, Destiny's Child, Britney Spears, Christina Aguilera, and Jennifer Lopez. Yet nothing in the class matched the ardent interest in Worldwide Wrestling Entertainment (WWE) sustained by Latino boys who identified both with the United States and Mexico, were bicultural and bilingual, and had strong family ties on both sides of the border. No other group of children stuck with a single interest for so long, and no other aspect of popular culture was favored for as long a time as the WWE. Unlike high-ticket consumer goods such as PlayStation machines, WWE merchandise was available in the neighborhood, as close as the corner convenience store. The WWE trading cards and magazines were coveted items, but for children lacking the pocket money to buy them, our computers with Internet access provided a highly desirable substitute, furnishing prized portraits of favorite wrestlers and action photos from the matches for their binders and home collections.

The WWE website functioned unusually well in providing the boys with what they wanted, and updating the content continually. While other websites functioned only intermittently on the lab's computers, and were plagued by technical difficulties, the WWE site had its audio streaming abilities up and running seamlessly in 2000, before music downloads were a commonplace of commercial Web design. Under the management of Vince McMahon the soap opera aspects of wrestling, especially backstage dramas and conflicts between managers and wrestlers, have been highlighted. WWE story lines about the wrestlers, their rivalries, and their struggles with management change on a weekly and sometimes daily basis, so the website always had something new for the boys. Video downloads are available and updated regularly, and instructions for downloading plug-ins to play them are easy to understand. The WWE offers video streams at variable frame rates designed to accommodate slower CPUs and dial-up Internet connections. The WWE is still a site that is unusually sensitive to the fact that not all Web surfers have the latest computers and connections. The WWE assiduously promotes license opportunities for merchandise and has worked hard to develop its tie-ins with popular music. Each wrestler has his own theme song, which, like a battle cry, announces his arrival in the arena. The music videos based on the theme and the wrestler play on huge screens at the events. CDs and DVDs of the wrestlers' themes are sold separately, but video and music are also available in abbreviated form as Web downloads. The music is usually stirring, with heavy-metal overtones. The students frequently sought out websites that related to their television viewing, but the WWE site was possibly the most closely connected to television and cable, both through its video streaming and its detailed scheduling information about local air times and channels. In addition, the site constantly promoted pay-per-view events, which the kids called "paper view"—referring to the paper bill that arrives in the mail following an order.

I was struck by the odd combination of children using Web browsers—a bold new use of technology in their lives—to pursue their interest in a very old-fashioned kind of entertainment. Later I learned that wrestling had a central role in twentieth-century Mexican culture, and that my students' fathers and other male (and sometimes female) relatives followed wrestling as a communal viewing experience. As Harold Rosen has remarked about the relationship between narrative and non-narrative discourses in children's writing, "there are always stories crying to be let out and meaning crying to be let in" (quoted in Willett, 2003, 12). Despite the fact that the WWE was always the boys' first choice of topic, I was slow to comprehend how meaningful wrestling was in their lives. This failure led me to view the boys' Web activities for some time as mundane or even compulsive, because I could not see that the material on WWE's sites was constantly changing. In failing to recognize the narrative complexity of wrestling and the multimedia blitz mastered by the WWE, the deeper implications of the boys' attachment to the WWE also eluded me.

In this chapter, WWE wrestlers and the stories that swirl around them are placed in a specific context of reception: an ethnically divided working-class community in southern California, where Latinos, African Americans, and African immigrants far outnumber Anglos. What did wrestling offer emotionally to working-class Latino students and what does it tell us about their negotiation of norms of masculinity? What happened when students translated this interest into writing for the newspaper? How can wrestling help students formulate—and even debate within the classroom—some of the ideological complexities that they face as poor children of color?

Anne Haas Dyson has documented the ways that boys of color in the Oakland elementary school she studied always chose to write about the media during open-writing periods. The boys had little desire to use writing for personal expression, or to establish their individuality or cleverness as students. Instead, they sought to use writing to make social connections with others. Dyson explains the children's use of popular media at school:

> in the third grade, working-class girls of color, unlike their female peers, also drew relative-ly often on the popular media, and working-class boys of color did so almost exclusively. . . . Their stories served as a means of displayed affiliation, of webs of connection, often under girded by the interplay of gender, race, and class. But in the public forum, the stories also generated ideological processes of value reflection and distortion and, thereby, new sorts of social processes. Among these processes were resisting by the excluded, distancing by the seemingly unimpressed, and, perhaps most important, negotiating among children who, after all, were without exception desirous of being powerful, respected members of a classroom community in which composing was a key medium for participation. (1997, 41)

Our goal in the newspaper class, like that of the teacher in Dyson's study, was to encourage the project of writing. If our official curriculum focused on learning research tactics on the Web and the genre of newspaper writing, the boys' writing about wrestling reflected their unofficial work of shoring up identity and creating a space in which they could feel powerful. As Dyson puts it, "Writing was not so much an expressive medium for individual souls as a tool for social beings whose major concerns were not learning to write" (42–43). In the classroom Dyson stud-ied, media materials range from the television show *X-Men*, to football broadcasts, to popular music. Although wrestling strays further beyond the bounds of the conventional curriculum than even the superhero texts Dyson studied, my students mastered a special kind of language and narrative in their wrestling stories, a lan-guage parallel to sports discourse. They competed to be the best at reproducing the official news and discourse of the WWE. To know the wrestlers' histories, to remember matches from years ago, to recall the precise terminology for special wrestling moves and types of matches (table, steel cage, tag team, bike chain), and to know about each of the programs and pay-per-view events (*Summer Slam*, *Wrestlemania*, *Raw*, *Smackdown*)—these skills were used to acquire status as fans within the group of male enthusiasts. While baseball, football, and basketball are now accepted as sports that can be used in the elementary school classroom to help motivate boys, especially in math, wrestling is denigrated as a sham, a farce, mere entertainment (as though the National Football League, the National Basketball Association, and Major League Baseball were not tainted by commercial entertainment concerns).

The boys enjoyed displaying their knowledge to me, too, and while I tried to be open-minded, I could barely comprehend what the boys were telling me about wrestling. In these conversations, they spoke very softly, and reeled off names of events, moves, and wrestlers and types of matches with which I was completely unfamiliar. So when one of the tutors helping out in the newspaper class revealed that he had been a high school wrestler (in the so-called Greco-Roman tradition practiced in high school sports programs), I asked him to talk to the boys about wrestling and help develop their article. The university student was a tall, blond surfer named Hunter Margolf, whose politics were decidedly liberal. Hunter was a student in my seminar where we read authors like Dyson on the topic of using and respecting children's interest in popular culture to encourage literacy activities. The students were charged with experimenting with a pedagogical method in which the children's interests would be respected and encouraged. Yet even with these explicit instructions, Hunter could not set aside his contempt for professional wrestling. He was intent on emphasizing the difference between Greco-Roman sport wrestling and professional wrestling, despite the boys' lack of interest in the former. The discussion between Hunter and two ten-year-old boys, Edwin and Andres, was audiotaped (a regular practice in the class when brainstorming ideas or conducting interviews as background for an article). The transcribed interview reveals the boys' repeated attempts to display their expertise on the subject and suggests its importance to them. Just before the excerpt quoted below, Edwin and Andres had been describing the various weapons wrestlers use: sledgehammers, baseball bats, folding metal chairs, and tables.

Hunter Margolf: Mm. So you think that's real? . . . Wrestling using the weapons and stuff?
Edwin: Probably.
Andres: Yeah.
HM: Kind of a different kind of wrestling, I guess. So what do you guys know about real wrestling? Like, the wrestling they do in college and stuff and the Olympics?
E: In Mexico.
HM: Do they have wrestling in Mexico, too?
E: Yeah, that's the original wrestling. 'Cause some Mexican wrestlers gone to the WCW.
HM: OK.
Kids: Yeah, there's a whole bunch of different types. Like there's the WCW, WEW, EWC, and WWF.
HM: So, which one do you guys like the best?
Kids: WWF.
HM: WWF's the best?
Kids: Yeah.

HM: Why's that? Is that where they use the most weapons?
Kids: Yeah, but that's where the almost all the best wrestlers are. And today, at 7 o'clock to 8 o'clock or 9 to 11, there's *Smackdown*.

Several things stand out about this section of the interview. Hunter tries to steer the conversation from the WWE (which the boys continue to refer to by its former initials, the WWF, Worldwide Wrestling Federation, an acronym abandoned to resolve a trademark dispute with the World Wildlife Fund) to the sanctioned versions of wrestling, exemplified by the Olympics. The boys proffer no interest whatsoever in the Olympics (see chapter 2 for other instances in which the teachers found the Olympics far more interesting than the children did). Instead, they politely dispute his claim that the Olympic matches constitute "real" wrestling. Instead they claim a Mexican origin for wrestling—referring to the form of wrestling known as *lucha libre* that originated in Mexico City in the 1940s and was also an important part of Mexican popular film (Levi 1999). The boys' attempt to counter Hunter's statement is barely acknowledged. Like so many Anglo teachers, Hunter is either afraid to display his ignorance of what the boys are referring to, or already discounts the information they are proffering. Instead he reiterates a somewhat ironic misinterpretation of what the boys are saying: that the best wrestling is WWE and is characterized by bloody battles with weaponry. Thus he returns to the widely publicized and not very interesting stereotypical position of boys being fascinated by violence (see Seiter 1999, chap 4). From there, it is a short step to the other familiar refrain of children being exploited by commercialization.

HM: Can you watch *Smackdown* tonight on regular TV?
Kids: Yeah.
HM: So, you guys have cable at home?
E: No.
HM: Yeah, neither do I.
A: I have, but I don't have pay-per-view.
HM: So, you can't watch the good matches unless you have cable, huh?
Kids: And pay-per-view.
HM: So even if you have cable, you still have to pay to watch some of these matches?
Kids: Um yeah.
HM: So, do your parents buy that for you sometimes?
A: No. Sometimes, I go to my friend's house and watch it there.
HM: So some of your friends' parents pay for pay-per-view?
A: They have the box.
HM: Oh, they have the box? [Pause.] Alright, so, you don't think the punches and stuff are for real?
Kids: No.
HM: Why not?
E: 'Cause they do like this . . . like that . . . [Physically demonstrating.]

HM: Yeah.

E: You see this picture? [Shows Hunter one of the images printed off the WWE website.] He doesn't even hit him.

HM: [In a puzzled tone of voice.] So you think the punches are fake but you don't think the baseball bats to heads are fake? Hmm. Why do you think they would make a baseball bat to head real but not the punches?

A: 'Cause it's more entertaining.

HM: Ah, okay. More entertaining. So, have you ever seen the wrestlers getting carried out by a stretcher and stuff?

A: Lots of times.

E: Yeah, lots of times . . . Triple H, and Jericho, 'cause he got hit in his face by the sledge-hammer. Oh, and Stone Cold Steve Austin, in the one that he got hurt he had to go for eight months to the hospital. And while he was over there the Rock . . . That's why the Rock's the champion right now . . . And he can't . . . He doesn't fight that much because his leg still hurts, his knee is broken, but he still wrestles. Now he doesn't fight that much because his leg still hurts.

HM: But he still wrestles?

Kids: Yeah!!

HM: Wow, what a guy.

In this section of the interview, Edwin and Andres continue to display their command of the topic: not only through their inventory of past injuries (which resemble a list of the kind of blue-collar injuries their relatives experience working construction) but also through their relating of show times, and the different formats in which wrestling appears—on broadcast network television, on cable channels, and on pay-per-view. The fact that the boys' fathers are also avid fans is not revealed here, although the boys are beginning to suggest this fact through their references to the neighbors—Hunter incorrectly assumes the boys are involved in a youth-only subculture that is too silly to be followed by adults. Hunter reveals that he does not have cable either, but his reasons are quite different from the boys'. Hunter wishes to be a firefighter in the national parks and believes in the value of spending as much time as possible out-of-doors; the boys' parents simply can't afford holidays that would require travel out of the urban area. The cost of wrestling fanship as reinvented by the WWE to include pay-per-view events makes it a constant target of unattainable longing for the boys. Yet Hunter cannot really follow what the boys are (accurately) conveying about the setup of wrestling on television, and keeps returning to the tired refrain of trying to force the boys to proclaim the fakeness of it all. Wrapping up the interview, Hunter makes one last stab at focusing the children's attention on legitimate, amateur wrestling as embodied in the Olympics:

HM: So, let's see. That's about it. You guys think it's pretty real except for the punches and the drops and stuff, huh? Did you watch any of the wrestling on TV during the Olympics? Like the Olympic wrestling?

Kids: No, like karate?

HM: Oh. Yeah, it was on TV a little bit. It was different from this, though. It was more like real wrestling, I guess.

E: Did they sell toys and stuff?

HM: Did they sell toys? Oh, no, no. The guys in the Olympics don't have all the promotional stuff like these guys. They just do it for uh . . .

A: To win?

HM: Yeah, it's not on TV and glamorized like that.

Two opposing views of the real are being expressed here. While conventional opinion would probably favor Hunter's definition of real sports as amateur events on the Olympics, the boys actually proffer a view of the real that is quite savvy about the business of wrestling. The boys' definitions resemble those of a network television executive, while Hunter's placement of the Olympics in a noncommercialized category of sport is somewhat naïve. The boys recognize that the greater the violence the more entertainment value the sport has, and that if something is really making it on television it will appear in the form of licensed merchandise—"toys and stuff"—which is exactly the kind of endorsed and licensed merchandise that many Olympians look forward to as soon as they have gone professional.

The first draft that Edwin and Andres wrote about wrestling touched on the racial and ethnic typecasting that is at the core of wrestling's appeal, while displaying their encyclopedic knowledge of the WWE. Still, it shows Hunter's influence in pushing the angle of the story toward the commonplace issue of whether it can be considered real. The first version of this story, titled "Is Wrestling Real or Fake?", went like this:

> Well, some kids say its real and others say its fake. But half of wrestling is fake like punches.
>
> But really studies show that its real. Sometimes people get abbsest [obsessed] with wrestling so they try to do the moves. But some times they don't successed [succeed]. Also, the wrestlers show bad examples to kids. And some wrestlers get so excited they say "I will face anyone 24 hours a day."
>
> There are 3 different kinds of leagues ECW, WWF, and WCW. ECW stands for Extreme Championship Wrestling. WCW stands for World Champion Wreasting, and WWF stands for World Wrestling Fedaration.
>
> Some wrestlers are from different states and countries.
>
> For example this tag team called Kai and Intai they're from Japan and they can't understand Engllish and they get translators for them and another wreastler named Essa Rios he's from Mexico and he also has a translator for him. Also some wrestlers are from Europe, like the bulldog. (Spelling and grammar errors in original)

Although most of the children acted in their daily lives as translators for adults between Spanish and English, this is one of the few mentions ever made of linguistic difference and translation in all of the children's writing over a four-year period. Wrestling was, in fact, the only topic in which the students in the class seemed to feel comfortable openly referring to language, ethnic, and cultural differences. The boys even switched to a special font associated with southern California Latino culture when writing about the WWE (see illustrations 7, 8). After the conversation with Hunter, the rewritten story developed in a different direction, more adjusted to journalistic standards. The topic shifted from the fans' fascination

Who is your favorite wresier out of hhh,HulkHogan and Kane?

What wresler is your favorite wwfsuperstar?

Why do you like wresling?

Since when have you bein watcing it?

Illustration 7. The boys used a font favored in Latino street culture to write about wrestling, one of the only topics where ethnic differences were openly expressed.

Illustration 8. Students staged a picture for their story using their wrestling action figures: the Rock action figure battling Stone Cold Steve Austin.

with wrestling to public opinion (gathered from class interviews) about wrestling. This choice was influenced not only by Hunter's conversation with the boys, but also by input from several other (Anglo and female) teaching assistants. When the student writers emphasized the potentially harmful effects of wrestling on children, the ethnicity of the wrestlers dropped entirely out of the story.

The new version was entitled "Wrestling Draws Fans, Opponents."

> Should children watch wrestling on TV?
>
> It depends who you ask.
>
> Children who were interviewed said they are happy with their choices about wrestling. A fourth grade boy said: "I think it's real 'cause of all the hitting and the cruel stuff."
>
> However, concerned adults including teachers and parents, questioned the violence involved in wrestling, such as the use of weapons and the destruction of furniture.
>
> Asked whether professional wrestling on TV is real or fake, kids said they think it's "mostly real." As proof, they said Owen Hart died because he was imitating another wrestler "Sting" by jumping from the top of the stadium. His death proved wrestling is not staged, they said.
>
> Those opposed to wrestling say children who are die-hard wrestling fans end up staying up late—maybe from 9 to 11 P.M. or every midnight, interfering with their rest for school or jobs. They say children pick up swear words from the "stars" they admire. Wrestling may teach negative values because families, including fathers and sons, fight each other. Children also may not realize that violence in general can be less than fun.
>
> "I think kids would get really hurt and they could start doing it (the wrestling on TV) to their brothers and sisters or siblings . . . ," said one girl, who had seen crime reports on the news.
>
> She added, worried, "Maybe they could start riots and they could go to their schools and they could just get a gun and shoot somebody." Also, other children said many wrestling fans rush through their homework so they can watch wrestling matches.
>
> This extends to the wrestling video games many children claim to enjoy. They are called "War Zone," "Attitude" or "Mayhem," and characters sometimes use slicers on each other or they hit each other with ladders or climb the ladders to jump on opponents in the games, as on the shows. The children say they pick their favorite wrestler on each game and may try to imitate their moves.
>
> The kids love the games but some adults are shocked. One baby-sitter stopped children from watching TV after a character threw a person into the corner of the ring, known as the turnbuckle.

Edwin and Andres conducted interviews with classmates revealing that girls as well as boys enjoyed wrestling, but that the girls were more adept at reproducing the teacherly, parental critique of the WWE. In later interviews, the Latina girls in the class displayed considerable knowledge of wrestling and spoke warmly of the experience of watching wrestling on TV with their fathers and brothers—although they often mentioned that their mothers did not approve. The girls maintained both positions simultaneously: enjoying it tremendously and voicing the presumed negative effects of the violence. But for all the participants, the significance of wrestling as a battle of types, representing class, gender, and ethnic positions, evaporated when the topic shifted to media effects. The second version of the wrestling story bore no trace of the ethnic typing, or of any linguistic or national differences so vividly represented in the WWE.

"The Original Wrestling": *Lucha Libre*

What were the boys referring to when they spoke of "the original wrestling" and "Mexican wrestling," in their conversation with Hunter? The affection of Latino students like Edwin and Andres for wrestling is no accident: it is firmly tied to their own class and ethnic identities, and shared with family members, friends, and neighbors. When Edwin and Andres mention real wrestling they mean the Mexican tradition of *lucha libre*. Started in Mexico City in the 1940s, wrestling enjoyed rapid success based on masked wrestling characters who combined a sort of Robin Hood social conscience with a mystical, even supernatural reliance on the sanctity of their mask. Initially event promoters merely imported professional wrestling acts from the United States, but *lucha libre* rapidly developed into a distinctive cultural form

In Mexico, the wrestlers, called *luchadores*, perform with tremendous acrobatic skill, moves that involve high-flying leaps into the ring and tremendous tumbling stunts. It is more athletic and more similar to gymnastics than the moves that are performed in the WWE; the *luchadores* themselves are much smaller and lighter than most WWE wrestlers, who appear to be heavy steroid users and typically weigh over 250 pounds. The worst thing that can happen to a *luchadore*, unlike the bloody pummeling of the WWE, is unmasking. Fan speculation on the identity of the wrestler behind the mask is an integral part of the experience.

More significant to interpreting the boys' reverence for wrestling is *lucha libre*'s association with working-class urban audiences and its themes of overcoming corrupt, authoritarian, and oppressive forces. Like U.S. professional wrestlers, *luchadores* come in two types: *tecnicos* (good guys or babyfaces) and *rudos* (bad guys). Anthropologist Heather Levi has traced the different strands of *lucha libre*: its media versions popular with the middle class, and its live versions popular with the working class. The former are closer to what is to be found in the WWE, while the latter contain a much stronger element of social protest and were therefore, like comic books, vulnerable to state censorship in Mexico:

> In the interactions between apparently suffering *tecnicos* and apparently underhanded *rudos*, *lucha libre* dramatized and parodied common understandings of the postrevolutionary system and their place within it. It reflected a political system in which people who

appear to be opponents are really working together. It paralleled an electoral system in which electioneering took place behind closed doors and elections ratified decisions that had already been made. Ongoing dramas in the ring demonstrated that loyalty to kin and friends is more important than ideology, and that arbiters of authority are not necessarily on the side of the honest and honorable. By its very name, *lucha libre* (which not only means "free wrestling" but "free struggle") resonated with the widely held and fundamental philosophy of the Mexican popular classes: life is struggle. This struggle was ritually enacted every week in the ring. (2001, 342)

Popular from B movies, television, and live events, this form of wrestling, as well as its social themes, was well known to the boys' fathers, and probably to the boys as well. *Lucha libre* exhibitions travel a circuit in the United States that includes San Diego, Los Angeles, and Houston, and northward to Canada as well. As one observer describes the matches, "Among the chaos you will find middle-aged laborers getting out their frustrations and forgetting about their troubles by hurling profanities at the villains. Their shouting might be overpowered by the elderly couples and children screaming unprintable chants at the top of their lungs just for the fun of it. Fans from every economic and social background come to witness the spectacle" (Brandt 2002, 9).

Lucha libre enjoys a certain status as legitimate urban proletariat culture, such that it has formed the basis for a museum exhibition, and several Mexican artists and intellectuals have immersed themselves in the culture (Levi 2001, 353). Figures of masked *luchadores* and the masks themselves are simultaneously being promoted now as urban kitsch, on the one hand, and as a segment of Latino youth culture on the other. For example, T-shirts printed with masks worn by *luchadores* can be purchased at the trendy Urban Outfitters stores, or the mall retailers of gothic wear, Hot Topic. Figurines and masks are available at headshops in urban centers. On the mass-market level, *luchadores* appear on some popular toys (the Magic Beanz collection) and star in their own Saturday-morning animated series on Kids WB—appearing as children—*Mucha Lucha*. The media conglomerate Warner Bros., eager to find ways to capture the Latino market, introduced *Mucha Lucha* in fall 2002 as an attempt to create a crossover viewership of teens, young adults, and anime fans. According to Rita Gonzalez of the Latin-American Cinemateca of Los Angeles, there is a cult following for *lucha libre* films: "I started noticing two or three years ago the popularity of Mexican wrestling with a non-Mexican audience, things like masked wrestling at clubs in Hollywood. Cine Luchador taps into the interests of young subcultures in L. A.; they're really drawn to the iconic image of the anonymous Mexican superhero" (Rommelman 2002, 36). There is a website devoted to the fans, Vivaluchalibre.net. There is a soundtrack CD for *Mucha Lucha*, which promotes some of the stars of the Warner Music label, such as Chicos de Barrio, Celso Pina, Tito Nieves, and Bacilos. None of the children ever mentioned *Mucha Lucha*, despite the fact that Kids WB ran special promotions, such as games in which kids could design their own masked wrestler. I was surprised to learn, when I asked the Washington students directly, that everyone was familiar with it because I had never noticed anyone in the class using the website. Some of the nonfans (of WWE) watched *Mucha Lucha* weekly, and easily reeled off the names of the three stars, Rikochet, Buena Girl, and the Flea. But the WWE fans shook their heads, saying *Mucha Lucha* was silly. They recognized that this was an outsider's version of "real wrestling"—the show's creators, Lilli Chin and Eddie Mort, are Australian—and found no appeal in the campiness of the animated series. Despite marketing the show to Latino children, WB executives were careful to emphasize the "universal appeal" of the cartoon: in one WB executive's words, "All I can say is, everybody gets it, *Mucha Lucha* is not going to be a barrier thing. It's like Taco Bell. Everyone knows what Taco Bell is" (Cobo 2002, 33). In other words, *Mucha Lucha* is on a par with the Chihuahua mascot. The boys rejected *Mucha Lucha* as inauthentic, and seemed to resent its humorous recasting of the *luchadores* as fighting for "Honor, Family, Tradition—and Donuts!"

Some critics have seen in the WWE themes about social injustice that resonate with the *lucha libre* tradition. Media scholar Henry Jenkins argues that WWE "stories hinge upon fantasies of upward mobility, yet ambition is just as often regarded in negative terms, as ultimately corrupting. . . . Virtue, in the WWF moral universe, is often defined by a willingness to temper ambition through personal loyalties, through affiliation with others, while vice comes from putting self-interest ahead of everything else" (1997, 59). The WWE is also exploiting interest in the *lucha libre* tradition through its promotion of the wrestler Rey Mysterio: the phrase *lucha libre* is never used in profiles of Mysterio found on the WWE websites, but the emphasis on his mask and the incessant repetition of the term "high flying" and references to his small size place him clearly in the Mexican tradition.

According to the *Smackdown* website, Mysterio lives in San Diego County (in National City, which is close to the border between California and Mexico) but has moved back and forth between the United States and Mexico. He speaks fluent Spanish and enjoys a huge Mexican (and Japanese) following. Mysterio was trained by his uncle in Mexico City. He is small (5 foot 3 inches, 140 pounds), but has the capacity to beat much larger opponents through his wrestling style:

> Blinding speed and springboard leaps allow Mysterio to run circles around most opponents with ease. . . . Acrobatic attacks and stunning counter moves spring off the ring ropes like an Olympic diver. . . . Fires off a relentless offense of high-risk maneuvers. . . . Atomic-charged wrestling artist that leaves them limp with his signature Huracanrana move. . . . Has been called the best pound-for-pound wrestler in the business. (*Smackdown* website)

Each aspect of the Mysterio character translates into merchandise and a stylistic association with southern California urban youth and Chicano styles. Mysterio wears a silver cross (a replica is available for sale through the website), enjoys hip-hop, uses as his signature "619," which is the area code for the southern half of San Diego stretching to the border. He began his career in Mexico, initially wore a mask in the United States, was unmasked in the ring in 1998, but has returned to wearing a mask.

Wrestling has traditionally been a cultural forum for slurs, stereotypes, entrenched ethnic rivalries, and bald racial prejudice. In her dissertation on professional wrestling in Los Angeles, folklorist Terry McNeill Saunders writes:

> In the 1960s and 1970s heel wrestlers thought nothing of insulting entire ethnic groups of fans to "draw heat," that is to make people very angry at them. One wrestler was famous for a routine where he speaks gibberish that sound like a foreign language to wrestlers who are (supposedly) non-English speakers and their fans in the audience. In New York, he would often grab the microphone from the announcer, and say that he wanted to speak a few words to the Puerto Rican fans. Then he would begin speaking with an inflection and accent that sounded vaguely like Spanish, enraging the Spanish-speaking fans in the audience, which was his intent all along. At the time, he was a "hell manager," and Pedro Morales was the babyface champion, beloved by all wrestling fans and the pride of the Latino community. He would say at the end of a show that next week he was going to bring his wrestler in and show all the "Mexican garbage truck drivers" that their champion was no good. (1998, 5–6)

In the WWE, racial insults have been reinvigorated and Rey Mysterio's narrative arcs typically take up themes of insult and respect. Jenkins argues that "The plots of wrestling cut close to the bone, inciting racial and class antagonisms that rarely surface this overtly elsewhere" (1997, 66). Mysterio figured prominently in a in a live WWE *Smackdown* event held in San Diego in 2003. The scripted conflict that was the centerpiece of the show was between Brock Lesnar—a heel represented as cheating for his title and in cahoots with the *Smackdown* manager, another bad guy—and Rey Mysterio.

Lesnar taunted the audience, in a manner reminiscent of Saunders's description of "drawing heat" from Latino audiences. Over and over again, he referred to the match as being held in Mexico, and when corrected by his manager, Lesnar claimed that he could not tell the difference, given the look of the audience. As the crowd booed him, he turned on them with the threat that if he did not get any respect from the audience he would get all of them deported. The audience, at least two thirds Latino, and about a third composed of children, held handmade signs with slogans like "619" against a background of the Mexican flag, or "Viva La Raza," or "Latino Heat." Then Mysterio appeared to tremendous cheering from the crowd, and challenged Lesnar to a match, saying he could not let him disrespect his people.

Wrestling and Symbolic Violence

Pierre Bourdieu coined the term *symbolic violence* to describe what takes place in the school system as well as in other social institutions that engage in the reification of hierarchical structures in society. Schools officially proclaim an egalitarian ideology, and there is much public proclamation of the idea that education is officially open to all, but in reality some groups hold a monopoly on its rewards. Symbolic violence is enacted in the distribution of life chances, and expectations:

> When powers are unequally distributed, the economic and social world presents itself not as a universe of possibles equally accessible to every possible subject—posts to be occupied, courses to be taken, markets to be won, goods to be consumed, properties to be exchanged—but rather as a signposted universe, full of injunctions and prohibitions, signs of appropriations and exclusion, obligatory routes or impassable barriers, and, in a word, profoundly differentiated, especially according to the degree to which it offers stable chances, capable of favouring and fulfilling stable expectations. (Bourdieu 2000, 225)

The wrestling fan activities involved an affirmation of their identities as Latinos—during a time when bilingual education at the school was essentially outlawed by the Blueprint for School Success and curricular reform mandates called for excising any Spanish from the classroom. Wrestling talk was usually done in Spanish. Whenever the WWE websites were pulled up, comments about the on-screen material and instructions to one another on how to successfully download video or audio files were in Spanish. These discussions of wrestling in Spanish continued over a period of years, and involved many different friendship circles of bilingual students.

These expressions of identity occurred in the midst of a schooling process characterized by Angela Valenzuela as "a powerful state-sanctioned instrument of cultural de-identification, or de-Mexicanization." Valenzuela notes that "'no Spanish' rules were common in US-Mexican schools in Texas throughout the 1970s, and California initiated a return to the notion of English as the gold standard of education." By the time the Blueprint for School Success was in place in 2000, the Washington principal was enforcing rules for no Spanish words in the classroom, and teachers were forbidden to translate English words on standardized tests for struggling students for whom Spanish is their first language after the passage of Proposition 187. This is just one of the ways that Latino "students' cultural identities are systematically derogated, and diminished. . . . Their proficiency in two languages is deemed a barrier to be eliminated. Their names are Anglicized or mispronounced" (Valenzuela 1999, 173–74). The boys' competence as translators between their teacher and their parents—who are trying to negotiate the school system when teachers know no Spanish and the paperwork from the school is rarely translated—seems a source of shame rather than accomplishment. Teachers and administrators openly display impatience with the students' parents, who are accustomed to radically different school cultures in Mexico.

The outward characteristics of the Latino students—compliant, polite, deferential—provided a stark contrast to the wrestlers they loved. Wrestling's appeal lies in its stark presentation of the unfair, the contest in which the fix is already in, and rewards are heaped on the egotistical (and Anglo) bullies of the world. To the Latino boys in my class, the school system must have looked like just such a rigged game, and the wrestlers perhaps expressed some of the boys' subterranean anger. Wrestling offers a universe inscribed with themes of social melodrama that bear a relation to conflicts encountered in their everyday lives—especially at school. Wrestling's value system mirrored what could be observed in the classroom community of Latino students, in which loyalty and friendship are rated more highly than getting ahead.

Latino boys were the most likely targets of ethnic slurs in the classroom. On several occasions, boys would be brought to tears after verbal battles in which they were attacked for their diets ("bean eaters"), weight, the poverty of their mothers (typical taunts being "you don't have any food at home" or "your mother's a maid") or the status of their fathers ("your dad's homeless" or an "illegal"). It was not only other children who created the hostile environment. Administrators freely disparaged Latino parents to anyone who would listen as failing to value education, the teachers complained that Latino families buy video games but not books, and make no effort to make school attendance a family priority. Janitors feel free to order the boys around. Even when I was present and had given permission for them to stay late or use the phone, one Anglo janitor shouted "Vamanos!" to clear the classrooms, when the boys were there after hours. In some ways the worst insults came in the form of "mock Spanish" of this kind (Hill 1998) instances where non-Spanish speakers inject Spanish words into their speech.

In my class, despite the fact that Valenzuela's book was assigned reading and we discussed the discrimination against Latinos in the public school system, I witnessed countless scenes in which university students who acted as teacher's aides presumed that the Latinos in the class were passive, slow-witted, and poor speakers of English. Indeed, the Latino boys differed greatly from the working-class Anglo boys, who constantly demanded adult attention, and frequently got in trouble, but were nevertheless treated as more intelligent. As Valenzuela points out, silence and politeness are often misinterpreted in school settings: "The 'politeness' and 'compliance' of immigrant youth follows logically from their lack of social power. Their 'politeness' is perhaps as much about deference as it is about powerlessness or an expression of their belief that they are not 'entitled' to openly defy school authority or assert their own vision of schooling" (1999, 140). The university students identified much more strongly with the interests of the Anglo students (video games, Neopets, Nickelodeon cartoons) and would often suggest new web-sites with additional information, or help the Anglo students navigate new sites. There was considerable tolerance, sympathy, or even support among the university students for Web research on more transient interests such as skateboarding, Heely skate shoes, Eminem, Harry Potter, punk wear, or the Insane Clown Posse. But wrestling was always sniggered at, as a bad joke. As Valenzuela notes about rebellious behavior among Texas high schoolers, "this is typical of the tendency in classrooms where acts of resistance among students who occupy a privileged posi-tion in the school population hierarchy and are in higher rungs of the curriculum (i.e., not remedial or ESL) are viewed more positively than when expressed by those of a lower status" (229).

Teachers' actions could bring the boys to tears. One day, after Edwin and Americo were selected to go to mandatory remedial summer school, I could see from their faces that something was terribly wrong. They entered the computer lab, but rather than race to the computers as they usually did, they quietly put their heads down on the desk and cried. The blame for the school's failure to deliver better test results is delivered to the Latino children and their parents. Edwin, Americo, Andres, and Carlos experience symbolic violence on a daily basis at school, through disenfranchisement and degradation. No wonder they gravitate toward the super-powerful and hypermasculine wrestlers as their idols.

The Rock

Rey Mysterio would seem like a natural candidate for the boys' favorite wrestler, and indeed boys like Edwin and Andres liked Rey Mysterio—but they vastly pre-ferred the Rock, whose implacable demeanor, enormous size, and insistence on get-ting respect offered fantasies of power that spoke to the boys' denigrated position and feelings of powerlessness at school. They especially enjoyed the power of the Rock vis-à-vis the leather-bound Aryan types such as the Undertaker, A-Train, and Brock Lesnar. While other wrestlers exerted their own fascination as figures of con-tempt or hatred, the Rock, whose real name is Dwayne Johnson, was always treat-ed with great reverence, his photos regularly downloaded and posted on children's walls at home, or made into special posters, or adorning their binders. The issue of ethnic identifications surfaced in the boy's first article about the Rock, compar-ing him to the white trash ultraviolent Stone Cold Steve Austin:

> Who is the best wrestler in the WWF? The Rock or Stone Cold Steve Austin? We attempt-ed to answer this question by exploring the backgrounds and fight experience of both of these outstanding wrestlers.

The Rock is called "The People's Champion," because he is considered the favorite among most WWF fans. The Rock is originally from Florida. Although wrestling wasn't his first career choice, he had dreams of becoming a professional football player, wrestling did run in his blood. The Rock's father, Rocky Johnson, was the first African-American to win the WWF Intercontinental title.

Stone Cold Steve Austin is also known as "The Texas Rattlesnake" to adoring fans. He grew up in Texas and wrestled with the WWF for many years before retiring due to a knee injury. During his retirement, Stone Cold passed his championship belt to The Rock. Steve Austin, who until his retirement was considered the most popular wrestler in the WWF, decided to give his title to The Rock stating that he would be the next star of the ring. Since then, Stone Cold has come out of retirement and has been currently wrestling for about four months.

Although between the two, The Rock was the last to hold the WWF championship belt, there is a controversy over who is actually the better wrestler.

The two well-known wrestlers have only met twice in the ring. The Rock lost to Stone Cold in their first fight, but defeated him in a rematch. Stone Cold has more experience and has wrestled longer than The Rock. In both matches between the two, Stone Cold wrestled with a serious knee injury. Despite the injury, he beat The Rock, but was unable to overcome his injury and lost in the rematch.

The Rock lost the championship belt on 10/22/00. He lost in a match against Kurt Angle. During the fight, Rikishi, another wrestler, entered the ring in order to help The Rock defeat Kurt. Unfortunately, his plan backfired when he [Rikishi] accidentally sat on The Rock, thinking it was Kurt Angle. This mistake gave Kurt the opportunity to pin The Rock and strip him of his championship belt.

Although neither The Rock or Stone Cold currently hold the championship title, both are still considered the best wrestlers, not to mention the most popular. Stone Cold lost his title to The Rock because of an injury, and The Rock lost his title to Kurt Angle because of a mistake. The world may never know who is truly the superior wrestler because both men refuse to fight each other and are considered best friends. Both also claim to be unconcerned with the title and simply compete for their fans.

The Rock was singled out as the figure of continual fascination for his expressions of ethnic pride combined with his hypermasculinity. His imposing physique and his classically handsome face are linked in his publicity materials to the constant theme of ethnic pride. In his autobiography, for example, the contents of which are endlessly reiterated on fan websites, the Rock continually references his racial identity, his respect for his African American father and Samoan grandfather, and his hatred of racism—especially racism coming from poor whites. The success of Dwayne Johnson in feature film production (*The Mummy Returns* (2001), *The Scorpion King* (2002), *The Rundown* (2003)) attests to some of his unique qualities as an actor and performer. These films often situate him in Latin American settings, giving him to a Hollywood version of a sort of pan-ethnic Latino identity. While Johnson's only ties to Latino culture are his marriage to a Cuban American, the children identified with him as a man of color, one who had no fear of telling off whites and who scrupulously maintained that honor and respect were the most important elements of his identity as a man of color.

Websites devoted to the Rock celebrate the highlights of his career, and these stories were memorized by the boys in my class. His career has included transitions from a babyface (wrestling's conventional term for a good guy) to a heel, or evil guy. But throughout these manifestations he has emphasized strong connections to family members, including his father (African American), mother and grandmother (Samoan), and wife (Cuban); an absolute contempt for racism of any kind; and a frank discussion of the ways his own life has been shaped by his racial and ethnic heritage. As the Rock tells the story of his father, Rocky Johnson:

> When my dad broke in, all of the top black wrestlers—not to knock them or anything—but they were all jive-talking caricatures. They'd come out to cut their promos and you'd swear they'd just stepped off the set of *Shaft* or *Superfly*. They'd eat watermelon on camera and do all sorts of degrading things, because that's what was expected of them. My father wouldn't do that. He was the first black wrestler to insist on being very intelligent in front of the camera. When he cut his promos, there would be no jive in his voice . . . and when he stepped through the ropes, there was no bullshit funk strut or anything like that. . . . He was a fearless man, a man who welcomed the chance to break down barriers. (Layden 2000, 10–11)

In describing his courtship of and marriage to his wife, Dany, he notes the obstacles stemming from her parents' prejudices: "Her parents were Cuban immigrants . . . they wanted their children to assimilate, to adapt and succeed. What they did not want was their daughter dating me, mainly because I was a person of color. So was Dany, of course, but that didn't seem to matter. I was half black, and that made me an unsuitable suitor" (117).

Finally, the Rock's performing career was characterized by a central conflict with Stone Cold Steve Austin. As a heel, the Rock would assail white crowds at WWE events with comments like: "There are twenty thousand pieces of trailer-park trash here tonight!" These are stories that speak very directly about the experience of racism and the complexities of negotiating identity as a man of color. The Rock offers the boys a figure of ethnic hypermasculinity, athleticism, and overt hostility against Anglos who insult him. At the same time, his star image emphasizes close ties to his extended family—his book is dedicated to his mother, "the strongest person I know." In short, he provides a figure of identification who posits a kind of utopian solution to the victimization the boys encounter at school.

Conclusion

Wrestling is not an ideal or even advantageous forum for children's feelings of powerlessness. It is, however, a form of popular culture that is charged with questions about race and ethnicity. In schools, "racial boundaries provoke the deepest questions of personal identity and social structure, as well as the deepest silences" (Jervis 1996, 548). For young students struggling with basic literacy, however, wrestling is as good as many other topics for learning about how to convey information and look up facts and details on the Internet. In criticizing the hostility to wrestling evidenced by my teaching assistants and by the teachers at the school, I do not mean to personally criticize these individuals. Rather I see this as symptomatic of a silencing of the interests of children of color. "All schools take refuge in silence about race to some degree, and many educators are novices at constructing multicultural, integrated school settings" (Jervis 1996, 548). Certainly the content of wrestling is understandably perceived as too challenging—and inappropriate—for the classroom. I am aware of the protection I enjoyed as a professor, author, and Ph.D. holder in the school environment: obviously teachers at Washington would be risking a lot to adopt the kinds of permissive policies I pursued, especially when they are so narrowly tied to the teaching of curricular standards. But teachers do not have to watch WWE matches on the classroom TV set to hold discussions of children's attachment to wrestlers, and the heavily stereotyped world of the WWE.

The denigrated status of wrestling makes Web-based fan activities ideal in some respects, since fans are always already cognizant of their lowly status. The Internet proved to be an exceptionally useful space for fan activity, given the disparaged nature of wrestling as a popular culture genre. As Saunders notes, "the Internet has reinforced the outsider/insider aspect of being a wrestling fan. No one in your family may understand why you like it, but there are hundreds of people to talk about it with you every day. . . . This need to defend oneself engenders a perverse pride in wrestling fans. This pride stems from persevering in the face of constant ridicule from spouses, family, friends, and co-workers" (1998, 209). For these children, critics of wrestling were primarily the adults they encountered at school, as well as some—although by no means all—of their female classmates. In some ways, their WWE fandom left the boys stranded in a netherworld—even if a high-tech, Internet-supported one—of the banal, and also the deeply sexist. Girls in the class sometimes tattled on wrestling fans, by taking me over to a computer screen to show me pictures of female wrestlers who appeared nearly topless, or to confiscate pages of the women wrestlers as "dirty pictures." Female wrestlers are photographed and their plot lines narrated in the style of soft pornography. Often the women's plot lines involve the juxtaposition of violence and seduction, such as when lovers, fathers, morality police, or jealous women interrupt stripteases. Matches are now routinely billed as "bra and panty matches" or, as one series billed it, "over the top, off with the top" matches. A wide range of organizations have protested the WWE content and its targeting of the young for vulgarity, sex, and excessive—and increasingly bloody—violence. The Parents Television Council (PTC) charges that the WWE deliberately promotes "disdain for authority figures, patriotism and religion" (quoted in Lowney 2003, 432).

When critics see only sex and violence in an area of popular media, any serious discussion of its representations of class and race are immediately closed off. While the media panic about wrestling is always framed in terms of the promotion of violence, the boys who followed it most avidly were also the ones who never got in school fights, never pushed and shoved, never knocked chairs over. As Toby Miller notes, discourse regarding media effects can be a "tool for encouraging concerns with social disorder to be leveled at texts of popular culture, rather than such issues as the availability of weaponry, the norms of straight white masculinity, the rule of capitalism, and the role of the state as an embodiment of supposedly legitimate violence" (2000, 45). Adults working with children need to listen especially hard and especially long to boys like Edwin and Andres, to seek out ways to interest them in writing, and to gain insight into the many facets of their cultural world that are invisible at school.

One of the initial goals of my project was to trace children's interests in popular culture through a study that was naturalistic and longitudinal. Critics of children's mass culture often ascribe significance to the success of children's commercial culture, without any context. When interviews are performed, they involve children who are already fans, leading to an overestimation of the extent or durability of the popularity. There are huge differences, however, between passing fads such as Powerpuff Girls or Sponge Bob and the longer-running, cross-generational, far deeper passion for wrestling that the Latino boys engaged in. While both kinds of media may evidence themselves through heavy cross-media promotion and an ever-expanding array of licensed merchandise at the moment of their peak interest, the difference between passing fads and more sustained involvement can be captured only through longitudinal ethnographic research. The case study of the WWE presented here suggests the extra effort that researchers must make to untangle the choices that children of color make among the media choices offered them, and the complexity of their motivations. At the very least, the WWE offers the boys a universe quite unlike the rest of the "World *White* Web." Teachers and researchers must make the extra effort to listen carefully to Latino children, while looking critically at those institutions of mass media and public school that do much to silence them.

chapter 5

Virtual Pets

Devouring the Children's Market

The phenomenon of Pokemon is the most powerful pop-cultural tsunami of its type yet to pour out of Japan and wash across the rest of the world. The reason for its success is its cross-media integration. Not just a computer game, not solely a film, more than a soft toy, a collectible that is also a card-trading game, TV series and cartoon book spin off, Pokemon is a commercial hydra, a polyhedral multifaceted simultaneous assault on global sensibilities.

—NICHOLAS FOULKES, *FINANCIAL TIMES* (2001, 24).

Pokemon and its online successor Neopets demonstrate the significance of the Internet in altering, stimulating, and broadening the market for children's toys and media. Pokemon created an entire genre of global children's media—one that is fueled by and sustained via the Web. Pokemon itself has faded—having graduated to what marketers call a *mature* license—but we can still expect that children ages seven to eleven will cycle through an interest in virtual pets and monsters, just as they cycle through interest in Barbies and remote-controlled vehicles. Just as that other notorious Japanese import Power Rangers—serialized, with dozens of changing cast members, like a soap opera, for well over a decade—is discovered anew by five-year-old boys long after its popularity has peaked with the older kids, very young children join the ranks of virtual pet fans each year. The Internet propelled the Pokemon craze by enlarging the market for video gamers to include girls, by expanding the fan base for Japanese popular culture products—especially anime—beyond a select group of aficionados to include a mass market in the United States and Europe. Pokemon, like its precursor Tamagotchi, proved interesting to girls as well as boys, thus offering a solution to the conundrum that had bothered the video game industry for decades—how to expand the video game market beyond boys. Pokemon marked a dramatic expansion of the potential for Japanese cultural goods with young audiences in the United States. This chapter examines the rise of virtual pets as a major category of children's entertainment, and the ways marketers employed the Internet as a new device for organizing the children's market.

Pokemon made more money than *Star Wars* or *Titanic*, and offered a new trajectory for children's licenses: video game to television to motion picture—with the Web functioning as a significant bridge between media. During 1999, the year of its peak popularity, video game software sales exceeded movie box office receipts for the first time in history (and this has been the case every year since). At the height of its popularity, Pokemon wrought havoc with brick-and-mortar toy stores, children's television producers, and the newly emerging online retailers who sought to target the children's market but failed to predict and supply for Pokemon's success. It set a new standard for children's television programming by proving that video games can provide sufficient narrative content for a series and interest kids both familiar and unfamiliar with the original video game. Pokemon started as a hand-held video game for the Game Boy platform, and mushroomed into a television series, a card game with collectible trading cards, several Nintendo 64 games, and three motion pictures. As a business venture, firms like Nintendo and the licensing agent 4 Kids Entertainment made a fortune on Pokemon, while other giants of children's entertainment, such as Mattel and Toys "R" Us, were seriously hurt by Pokemon's decimating effect on the sale of traditional "evergreen" licenses such as Barbie and Hot Wheels.

Pokemon set the stage for the development of the online role-playing game Neopets, which took advertising to children to an entirely new level. Neopets is a role-playing game in which players own creatures that are part Tamagotchi and part Pokemon. Each pet has a cute name and body (owing much to anime, although the original material was designed by British college students) and comes with a variable set of traits that range from their fighting abilities to health and personality. There are forty-four different species of pets available, including rabbits, cats, and dinosaurs. Closely modeled after Pokemon are strategies for cross-gender appeal: Neopets players may gravitate toward caretaking and adoption or abandonment and neglect, pleasing, friendly behaviors or mean and nasty ones, and can accumulate points through combat or through shopping. Neopets is a Western imitation of the style of anime and tone of Pokemon, sold back to Asia via the Web, eliminating the television series as an intermediary phase.

Defying Industry Wisdom

A rapid increase in domestic access to the Internet in Europe and the United States (especially among families with disposable income) coincided with the Pokemon phenomenon in the 1990s. When interested fans searched for cards in stores, only to be frustrated by the lack of merchandise, they could turn to the Web to read up on Pokemon, exchange game tips, or view cards for auction on eBay. Amateur web-sites sprung up rapidly, linking fans, game players, and card traders globally. When AOL bought Time Warner, it moved quickly to capitalize on such labors of love by amateur fans and Web designers, and offered to sponsor dozens of Pokemon sites on AOL, or buy them out completely. Thus corporate promotion could appear under the guise of the homespun fan page.

During the relatively long period of time in which merchandise in toy stores was scarce and information about Pokemon was in huge demand, eBay drew thou-sands of new customers through its offerings of Pokemon cards for sale. The auc-tion site was founded to cater to private collectors: it advertises itself as the best place to trade and to meet and talk to others who share the same interests. The founder claims to have invented the site to provide an access for one of her pas-sions—collecting Pez dispensers. But this charming origin story (cited in the busi-ness press as one of the most successful public relations spins of recent years) masks the fact that the CEO was a very experienced executive, with a long histo-ry of brand management at Proctor and Gamble. Pokemon card trading was espe-cially well suited to online auctioning. EBay has had one powerful advantage over other Internet firms: it makes its money entirely on transaction fees and never stocks merchandise. This unique characteristic has made it the darling of Wall Street—one of the few firms to survive the dot-com free-fall of 2001 and consis-tently turn a profit. At the peak of Pokemon popularity, as many as thirty thousand Pokemon items were available on eBay for auction—many with color photos of the cards or toys available. So while an avid fan might have had trouble finding the latest imports from Japan, thousands of cards could be viewed online, and Pokemon fans could engage in long e-mail exchanges with eBay sellers about their goods. Although eBay now sells thousands of goods from major retail chains such as Sears, during the Pokemon craze the site was still dominated by individual fans and collectors.

In February 1999, the search for Pokemon merchandise also helped make e-Toys the fourth most visited online retailer. Parents, frazzled from trying to buy gifts for their children by anticipating arrivals of new stock, turned to the Web (Dodge 1999). While Toys "R" Us lagged far behind in ordering Pokemon products, e-Toys offered Japanese-language Pokemon cards, deluxe sets, and photos of cards not yet released in the United States. The failure to supply Pokemon merchandise was one factor in making Christmas 1999 a disastrous year for Toysrus.com, which led to a total reorganization of the company and the handing over of its website management to Amazon.com. By March 2001, e-Toys had filed for bankruptcy, citing rising costs and competition from Wal-Mart. There have been no leaders in the toy field since Pokemon of the kind that can build traffic in stores and drive parents as well as children into shopping frenzies, and the novelty of ordering goods online had worn off.

Of all the U.S. companies profiting from the Pokemon craze, the clear winner was 4 Kids Entertainment, the handler of the license. The success of the property catapulted 4 Kids to the top of the list of the fastest-growing Fortune 500 companies. 4 Kids experienced a 400 percent increase in revenues in one year. Besides selling the license to toy manufacturers, 4 Kids handled TV sales of the property worldwide to sixty countries, where the program topped the prime-time slots in Germany, France, the United Kingdom, Italy, and Spain. 4 Kids owned nothing more than a share of the intellectual property, and tried neither to produce advertisements, nor to sell merchandise or make animated cartoons.

Besides providing dazzling profits for Nintendo, Pokemon played a role in forcing its major rival, Sega, to get out of the hardware business completely—abandoning its Dreamcast platform (the major contender with Nintendo 64). Thus the tiny Pokemon helped Nintendo eliminate one of its major competitors in the hardware market—and video game hardware accounts for about 50 percent of Nintendo's profits. Over three quarters—76 percent—of its profits derive from overseas sales, which means that Nintendo tends to benefit from a weak yen (Briefing Books 2001). Game Boy units proved so robust that no one in the industry expects that platform to become obsolete. As explained in a Japanese business news article from March 2001, mobile game equipment was faring better than stationary, fixed game equipment. It is now considered the standard in "anywhere, anytime" video gaming, and the lower-cost equipment—attractive because it can be used by kids in the car, at the grocery store while kids wait for their parents, or at the airport—is considered a form that will never go away (Hattori 2001). In 2001, Nintendo introduced Game Boy Advance in the United States, with 32-bit color and a wider screen, but designed it so that it still plays all Game Boy cartridges. In 2004, Sony backed down from its insistence on Playstation and released a hand-held device similar to Game Boy.

Turning Japanese: The Appeal of Pets and Monsters

In the United States, Pokemon play was originally the domain of adolescent boys but rapidly spread to elementary-school children and to preschoolers. Most unusual for the toys industry—normally rigidly segmented by gender (Seiter 1993)—was the fact that girls as well as boys became avid fans. On street corners and playgrounds, girls and boys crossed their separate media and toy worlds to trade cards, play Pokemon games, and play video games.

No one anticipated Pokemon's diverse, spectacular appeal. Pokemon did not plan a marketing scheme to achieve its broad appeal across gender, age, and ethnicity—much of it happened through word of mouth fueled by the Internet. Pokemon marketing originally targeted the traditional base of video game enthusiasts—in the United States, Anglo adolescent and young male adults. In the month prior to Pokemon's first release on Game Boy, over 1 million videotapes were mailed to video game enthusiasts explaining Pokemon. Many industry professionals were doubtful about its potential beyond this group: it seemed odd and too Japanese. Video game enthusiasts are often already anime fans and are familiar (from earlier video games and movies) with the genre of role-playing games, the visual conventions for representing emotion found in anime, and Japanese cultural elements contained in the games. Japanese game designer Satoshi Tajiri based Pokemon on childhood memories of collecting bugs in jars and wishing he could make them fight like the monsters in his favorite science fiction movies. Tajiri spent six years developing the game, which was intricate enough that it was impossible to translate directly to English. In an interview Tajiri said, "I was told that his kind of thing would never appeal to American audiences. Because the characters are in a Japanese style, you cannot sell them to Americans" (*Electronic Gaming Monthly* 1999, 163).

Pokemon tapped into the global market for cute animals with huge eyes begun by Sanrio's Hello Kitty; the style is identified in Japan by the term *kawaii*, which suggests cuteness and an association with girls' culture. Under Sanrio's leadership, *kawaii* has become a highly lucrative cultural export for Japan. This cuteness—combined with an emotional appeal to nurturing—was the big draw for girls to virtual pets (Bloch and Lemish 1999). Being a good Pokemon trainer (or Tamagotchi or Neopet owner) is a matter of avoiding favoritism and giving proper pet care. Players must provide growth experiences for their virtual pets, following the developmental model of child rearing that the children themselves are subjected to. So if Pokemon trainers are something like the classically masculine Samurai warriors wandering the countryside in search of battles, they are also very domesticated. Besides battling, trainers must engage in such feminine pastimes as checking in at the mall or the gym to prepare for the next encounter, or dropping off their Pokemon at day care if they can afford it.

The caretaking model has proven to be ingeniously adaptable to a digital realm, where the vagaries of sickness and accident can be programmed into the game and the efforts of the individual player/custodian can be precisely tabulated and recorded over weeks or months of intermittent play. The caretaking function that makes Pokemon and Neopets more popular with girls than previous video games has its origins in the Tamagotchi. Rather than springing from role-playing games or video game genres, Tamagotchi has an origin story as charming as Pokemon's. In this case a Japanese mother wanted to satisfy her children's yearning for a pet. Real animals are difficult to maintain in the restricted spaces of Japanese residences, but virtual pets so small they could be held in the hand and transported everywhere seemed a perfect answer (Bloch and Lemish 1999). The digital devices re-created the life cycle of the pet—hatching from an egg (*Tamagotchi* is a diminutive form of the Japanese word for egg) to growing to adulthood, with sleeping, washing, eating, and exercising required. In 1997, the toys, produced by the Bandai Corporation (makers of Power Rangers action figures), experienced a meteoric rise to global demand, followed by a precipitous fall, with Tamagotchis and their imitators overstocked on toy and retail chain shelves everywhere. Pokemon and Neopets followed closely on the success of Tamagotchi with one major alteration: the pets never die. On Pokemon they may weaken or be traded, and on Neopets they can be abandoned and re-adopted, but death is not a possibility. Neopets exemplifies a Western imitation of *kawaii*—produced in the West, and sold back to Asia.

Pokemon did not crack the girls' market by offering a host of positive female role models. In fact its success belies the conventional marketing wisdom that girls like girl characters with which they can identify. While sometimes portrayed as a serious trainer, Misty falls into the role of infatuated schoolgirl—even groupie—in her relationship to Professor Oak. The series and the video game do offer a range of female characters in the roles of "gym leaders" in the various cities where Ash travels seeking new experiences and battles in which to train his Pokemon. Nor did Pokemon offer politically correct images of sexuality. In Japan, mothers do not object to the bare breasts and crotch shots of anime or of the manga comic books so hugely popular with adults and children, considering them a harmless escape (Allison 2000). Expecting that this would not be tolerated, certain aspects of Pokemon were altered for the less permissive U.S. market: for example, the line in the Japanese version of the theme song that says Pokemon can go anywhere—even up girls' skirts—was eliminated.

Nintendo hired 4 Kids Entertainment to translate the original 150 pocket monsters into names that could be comprehensible to Americans and trademarked in the U.S. Koichi Iwabuchi has identified this "deodorizing" of Japanese products for the American market as an increasingly deliberate strategy of Japanese companies seeking global exports (2004). Still, many Japanese cultural elements—some of them quite foreign to U.S. audiences—were retained. The child characters are devoted to a life of study and training, which is declared in the show's theme song with the lyric "I want to be the very best, like no one ever was." The television series celebrates holidays unknown to U.S. audiences, such as Children's Day. Minor characters are often situated in intense emotional situations related to studying for exams, wishing not to disappoint one's parents, or gaining acceptance to the best schools. Pokemon trainers also encounter an array of elements, such as spirits and ghosts, with decidedly animistic overtones. Nevertheless, Pokemon was single-handedly responsible for catapulting Kids WB (the Warner Bros. block of morning and afternoon programming) to the top of the cable ratings. When episodes were first dubbed into English (and Japanese characters removed from the visuals—airbrushed out of the animation) it was done at the expense of the licensing agent (4 Kids Entertainment), which offered the Pokemon episodes to cable under a barter agreement. In other words, none of the major players in children's television saw potential in Pokemon because of its overly Japanese "cultural odor" (Iwabuchi 2004).

Pokemon falls into the category of role-playing games (RPGs), a genre of enduring popularity in Japan, which had never before succeeded on a broad scale with American audiences. In the United States, Dungeons and Dragons and Magic: The Gathering were the only successful RPGs before Pokemon. These were only popular with a restricted group of fans, a group somewhat overlapping hard-core video gamers and anime fans. Dungeons and Dragons and Magic were unpalatable—repellent even—to mainstream U.S. suburban culture. The games themselves bore gothic connotations and were mockingly typed as "men in tights" stories, and the fans were stereotyped as creepy, unathletic, unpopular, and unattractive teenage boys. The Christian Right vehemently condemned the games as Satanic. By contrast, RPGs are openly accepted in Japan and popular with a diverse audience. American parents tend to see a child's (and it is usually a boy's) immersion in the kind of fantasy world represented by RPGs as disturbing, abnormal, and antisocial; Japanese adults see this as a harmless, fun distraction from studies (Allison 2000). Because adults as well as children are regular readers of manga, or comic books, fantasy is not considered, first and foremost, a children's genre.

It makes sense that Pokemon broke RPGs through to the U.S. audience, because it avoids the more menacing associations of Dungeons and Magic. What proved so absorbing about Pokemon was that it offered a multilayered, intricately structured world—something possible only within an RPG. The card game was so complex that few of the collectors managed to master it, and stores like Wizards of the Coast hosted frequent demonstrations by expert game players to help familiarize audiences with the nuances of play. Pokemon was much more than just collecting one cute pet after another, first 150 of them, then another set of 150 (although this was itself an audacious move in terms of prospective licensing profits). It was by all accounts a highly original, vividly imagined, and painstakingly realized fantasy environment.

Pets on the Web

During the same period of time in which Pokemon wrought havoc in the toy and television industries, producers of new media for children experienced their own set of failures. While many optimistic producers had envisioned a future in which television sets and video games would be abandoned by children for the Web and interactive CD-ROMs, it became rapidly clear that the dominant children's media industries would remain dominant on the Web. Huge mergers in the industry, such as AOL–Time Warner, helped to bolster the presence of children's television on the Web. But it rapidly became clear that, just as children have proven fickle as TV viewers, they were fickle on the Web. Children did not click on banner advertisements (neither did adults); they easily avoided standard advertising practices on the Web. Subscription services for children failed, despite many efforts to cash in on parental anxiety over letting their kids loose on the Web. Only AOL remains standing as a viable ISP that advertises to parents with claims that it will monitor children's activities on the Web and block pornographic sites, and AOL rapidly lost market share to DSL services between 2002 and 2004. The futility of its blocking efforts became clear when it was revealed that one of its employees had sold thousands of email addresses to spammers.

The answer to the problem of children's wandering attention was online games. As Dan Schiller explains, "Games were turned to because of their potential for interactivity, or what they prefer to call 'impulse interactivity'" (2000, 93). Today it is recognized that the "stickiest" websites are game sites. Games not only capture the audience but also capture sponsors, "who can lard them in all sorts of creative ways with product mentions and demonstration" (94). These online games also offer more aggressive ways to encourage repetitive play. The success of Neopets must be appreciated in this climate of business opinion in which entrepreneurs have despaired of capturing children as a market on the Web, and hundreds of websites have gone under since the mid-1990s—a process accelerating by the ongoing economic crash. The business press repeatedly praises Neopets.com as a model for future efforts to turn a profit from the Web. Neopets constitutes one of the most successful—one of the stickiest—websites yet designed. According to a 2002 study, 30 percent of tweens, eight- to twelve-year-olds, rank Neopets.com as their top site, followed by Warner Bros. and the Cartoon Network. In Canada it is the number one destination for male teens, who spend an average of 576 minutes a month on the site. (It is the second favorite site for girls after MSN Messenger.) In June 2002 the community of Neopets owners was estimated at 32 million; 39 percent of members are twelve or under, 40 percent are thirteen to seventeen, and 21 percent are over eighteen (Dennis 2002).

Neopets stands out as a further example of the evolution of the children's market away from a total dependence on TV series and advertising. (No television cartoon exists for the concept, although Viacom has contracted to make video games featuring the characters.) Neopets was first launched commercially in February 2000 (Weingarten 2002, D5) The geography of Neopets.com is dense and complex, scaffolded, like video games themselves: qualitatively different levels of access reward the investment of many hours of play. A staff of thirty game designers, working in Glendale, California, update and monitor the content each day. As an online environment, play at Neopets includes such proven favorite activities on the Web as gambling, simulation games, and competition against anonymous online gamers. When players log on to Neopets, they navigate a series of worlds embedded in the game where they can gain points to expand their collection of pets, buy food or services for their pets with points, or engage in combat with other pets. The worlds resemble a Tolkien story crossed with a theme park: the Lost Desert, Tyrannia, Faerieland, Terror Mountain, Mystery Island, Virtupets Space Station, Haunted Woods. In each world there are stores and games. The stores stock food and accessories that can be acquired through purchase with Neopoints, or can be bid on, like an eBay auction. Players themselves host the stores, and much of the fascination with the game has to do with trading assets player to player. Players can establish their own online store offering accessories and treats for their Neopets. These take the form of pet pets, or toys and games for their pets, or even books to read to them. At this much deeper level of play, product placement occurs. For example, Neopet owners can acquire Capri Sun drinks to feed their pets, or Hot Wheels cars or Mattel's Diva Starz dolls for them to play with.

Neopets offers games in the form of puzzles, action, or a category called "Luck/Chance." The aspect of Neopets closest to Pokemon—the old-fashioned way of winning points in video games—is participation in battle sites, where a Neopet is placed in competition against a single combatant, and the player chooses from a store of weapons, moves, and skills to attempt to win the game. For players uninterested in combat (and for those lacking the knowledge of the lore and strategy to succeed in battle), points can be earned by viewing advertising—usually these activities gain much larger numbers of points than can be had playing a game. The simplest form is visiting websites—for the Cartoon Network, for recently released films, or for various retail items and services, such as comparison shopping portals. So for example, a visit to the *Spy Kids* online theater might gain a player 250 points, but a visit to a comparison shopping website, in which one required task is to get prices for three different electronic devices (digital cameras, MP3 players, and DVD players, for example) can earn a player 1,650 points.

Exercises that are more lucrative require divulging more data about oneself. Neopets.com rewards those who disclose personal information (address, zip code, telephone number, e-mail address) and one's consumer preferences (answers to polls, choices of name and color and style, desirable sweepstakes prizes). Surveys are offered on a regular basis, collecting information about age, gender, location, use of the Internet, and frequency of candy consumption. In one 2002 survey worth 800 points, kids were asked to name their favorite lollipops, select from a list of brands those they had heard of, and then describe "how they would feel" about having a new kind of Starburst lollipop on Neopets. At the end of the survey participants were thanked "for helping make Neopets a better place." As a business enterprise, Neopets is selling information about the children and young adults who are its fans, rather than selling a media product itself. The more information a user surrenders, the better she or he is positioned to play the game, publish his fan drawings, stories, or poems, and gain Neopoints.

Neopets capitalizes on every genre of fan art and fiction popularized around television in the last twenty-five years and widely disseminated through the Internet (Seiter 1999). (See illustrations 9, 10.) On the Web, these varieties of fan

Illustration 9. A student's drawing of her favorite Neopet; the website offers opportunities to post fan art.

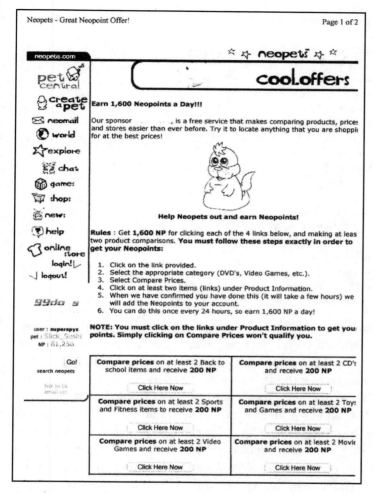

Illustration 10. Neopets awards points for shopping, as well as gaming.

expression are accumulated and sent out to audiences far larger than that of traditional fan associations, newsletters, and 'zines. Neopets publishes a weekly online newspaper, which offers opportunities to publish short stories, drawings and paintings, poetry, news stories, strategy tips, and personal testimonials on the Web. While to a nonfan these look like kitsch, and many are blatantly juvenile, they are carefully blended with the site's own official publications—such as the *Neopian Times* and its promotional materials describing the history of Neopia. A fan's poem describes the emotional identification with virtual pets, and exemplifies the way that Neopets have fit well into a girl's fantasy genre:

The Lonely Aisha—By venuslovegoddess16
As you walk through
The rubbish littered street
You see her up ahead
There she lies, sad and dirty
With nothing but papers for a bed
She doesn't wag her tail
Or run to you with glee
For in her eyes
There's a glint of sadness
Which we pretend we can't see . . .
You take from your pocket
A collar and lead
And when she looks at you with love
You know you've helped a pet in need
So always adopt
If you can afford the cost
Because they're the pets
That need you most
The pets that are sad and lost

Other aspects of Neopian society borrow from more sophisticated simulation games and target an older audience: Neopia has an inflation rate, a stock market, and possibilities for fraud. Pets hold bank accounts and can earn interest or buy stocks. Similar to accounts of the Pokemon economy, one enthusiastic reporter dubbed Neopets "a playground for the junior capitalist. Think Tamoagotchi . . . Monopoly, NASDAQ and eBay all rolled into one" (Tam 2001, 83). Fraudulent activities might be the subject of an editorial in the newspaper, or discussed in an area called "soap box," where fans express opinions about the running of Neopia and the fairness of play—hacking others' accounts and stealing prized collections have been common problems in the Neopet world. The third way of gaining Neopoints is through gambling, or games identified as "Luck/Chance." All kinds of games of chance have been adapted for Neopia: poker, blackjack, keno, bingo, roulette, slots, lottery. Neopets—like hundreds of other Web promotions—introduces children to the pleasures of gambling long before their legal entry to a Las Vegas casino. And yet when they go to Las Vegas, they are likely to visit gaming palaces owned by Nintendo, hotels owned by the parent or partner companies of the Cartoon Network, and casinos operated by the same media conglomerates that dominate the children's industry. Eased restrictions on gambling in the United States have produced one of the largest growth sectors of the U.S. entertainment industry. Their presence is very visible on the Web—pop-up banners for card play and casino games appear on the children's screens every time they search the Web in our computer lab. To see Neopets' development of this type of game playing on its site is to watch the recruitment of the next generation of gamblers—one of the "killer apps" of the Internet—as much as it is to see kids recruited for watching cartoons on TV.

Neopets strives to add the same compulsion to play that has characterized video games and has been such a concern for adults. Like Pokemon gold and silver versions, Neopets undergo changes while a player is logged off—including sickening from neglect and becoming obstinate about following commands in battle. New materials and new games are posted daily, and players soon recognize that playing at less popular times of day will be rewarded with greater ease in gaining points, or better access to specially prized areas in the website. Thus, Neopets rewards daily, obsessive play.

Neopets is also an experiment in a newer form of risk-averse toy marketing, where the lag between pent-up demand for merchandise and the delivery of goods does not destroy potential profitability. This is the lesson learned from the Pokemon phenomenon, in which the firm that made the most money was 4 Kids Entertainment while the brick-and-mortar stores, or Internet catalogs, or manufacturers, got left holding the bag with way too much merchandise timed too late. Neopets originally offered only a very small amount of licensed merchandise—primarily in the form of T-shirts. Next it used the website to test-market a wider range of product offers. Through Web announcements and e-mails sent directly to players, the first toy sales were staged as exciting events. Neopets merchandise was offered for only a few days at a time, in a few strategic locations—close to the home base of Neopets in Glendale, California, a suburb of Los Angeles. For several weeks, the website announced the following event: "For one day, on *November the 10th*, Neopets will have real world merchandise on sale as a test. Boohoo! There will be Neopets stickers, plushies, caps and more . . . available *all day* at the following three Southern California locations" (emphasis in original). Later that month, Neopets sent e-mails to anyone who had clicked through on the store locations:

> Our first test was a complete success!! Thanks to everyone who was able to show up and support us that day. We are allowing these stores to put any remaining Neopets merchandise back on sale for the next couple of weeks until it is sold out. Plenty of plush and stickers, some jewelry and notepads! There will not be any staff there handing out free rare item codes, but there are rare item codes in some of the merchandise.

This makes possible the delay of mass production until the retail goods are proven winners. The data gained from the Web helps to gauge the potential market—so a limited roll-out of merchandise can still be widely publicized, and demand can be built slowly. It is an approximation of bringing "just-in-time" production techniques to the realm of toy sales. When toy sales become a fan event, rather than everyday shopping at Toys "R" Us, there is no danger of price cutting, a practice that has become more widespread with the deepening U.S. recession and has undercut toy industry profits overall. The strategy also works hard to keep the atmosphere of a casual fan club ("thanks to everyone who was able to show up and support us that day") while tying a visit to the retail store to enhanced gaming ability (such as opportunities to secure rare item codes for collection in the virtual environment). In the summer of 2002, Neopets announced its retail agreement with two stores targeting girls that are commonly found in malls across the United States: the Limited, Too, and Claire's. E-mails announce to customers the release of each new plush toy, which are dribbled out a few species at a time (one recent e-mail sent to all members announced the availability of new plush items with the warning "if you are going to want one of these for a holiday gift, this delivery is all that will be available between now and the end of the year."

Neopets' marketing techniques are the brainchild of chief marketing executive Peter Green, who was the creator of *Fox Kids Club Magazine*, and before that worked on licensing campaigns for the Disney feature films *The Lion King*, *Aladdin*, and *Pocahontas*. Green has signed Viacom to begin producing video games and books. Meanwhile, *Advertising Age* commissioned Neopets to carry out a study of the Internet compared to television and impressions of advertising. They have also trademarked the concept of "Immersive Advertising." Neopets boasts Kraft foods, Nabisco, Proctor and Gamble, Disney, Universal, and Warner Bros. as clients, and calls itself the "leading online global youth network." For its clients, Neopets offers extensive tracking of brand awareness and adoption of its virtual premiums, through pre- and postcampaign market research. A pleased Mattel executive remarks, "It's usually impossible to measure the exact effect of online initiatives but [those] data show you exactly how your brand is doing" (PR Newswire 2002). In it capacity as market researcher, Neopets projects that young people spend 12.1 hours on the Internet per week as compared to 7.5 hours watching television, and that the Internet is preferred as a more engaging and less passive experience than television, but that children prefer advertising messages on television.

Ralph Nader's commercial alert attacked Neopets for offensive tactics, but Neopets responds to critics with the claims that its techniques are less annoying than banner advertising (which no one will pay for anyway), and that advertising represents "less than 1 percent of the site's content." Green explicitly compares online games favorably to TV viewing: "Look at Saturday morning television, the show lasts for 22 minutes and eight minutes of that is commercials. Our appeal is that there is so much content and it's all free—but we stay in business because of our advertisers" (PR Newswire 2002, 1).

A company press release quotes one fan's response to criticisms of Immersive Advertising: "This is better than blatant and ugly banner advertising!" The website includes a persuasive statement to parents about "why it is safe to go on Neopets." The emphasis is on Neopets not giving out your child's e-mail address to others. The barriers that do exist, however, are easily overcome by children, as one of my ten-year-old students cheerfully reported in a newsletter article: "You can join the many users by going to Neopets.com and signing up. You'll probably need to think of a username, password and your email address (don't worry, it's not going to be shared). Now if you want be able to neomail you should sign up for that, but be sure you give your age as over 18. Now get to your computer and sign up and start having fun!" As legal scholar Marc Rotenberg pointed out about the Consumer Privacy Protection Act of 2002, the only provisions that exist "require a company to adopt a privacy policy that can say virtually anything and can be changed at any point in time to say anything else . . . there is an interesting section that attempts to limit the sale of personal data to third parties, but this provision is easy to defeat by simply offering the consumer a benefit, such as the service originally sought." Rotenberg dubs the ease of shifting privacy policies the "Now you see it, now you don't" policy provision (2002).

In June 2002, Neopets announced that the company was considering moving its Neopets Asia division to Singapore to take advantage of tax incentives and the availability of animation workers, pending the outcome of talks with the economic development board. Singapore Technologies has a $5 million stake in the L.A.–based Neopets. In Singapore, it is estimated that over 600,000 young people use Neopets, which they see as an advanced version of Tamagotchi. Players are called Neoparents, making the nurturing aspect of the game more blatant. An after-school tutoring company has partnered with Neopets to offer premiums and rewards for good scores on the monthly assessment tests in English, math, Chinese, and science in the form of pet pets, treats, and gifts for their Neopets. The child who scores well is given a special code, which when entered on the site earns the special prize. Rare prizes are made available to the enrolled students, who can then barter them on Neopets for points. The idea is that children will then spend less time on the Neopet Internet site striving to gain points, and instead shortcut the process through gaining good grades. As reported in the business press, the idea came to the managing director of MicroMax when he felt that his own daughters were neglecting their homework for Neopets. His daughters wept when he threatened to put their Neopets up for adoption at the virtual pound, so instead he decided to learn about the game and try to use it to motivate them. "I even had my daughters teach me about it—how to make them smart, keep them healthy and how to challenge other pets. It's pretty complicated, you know!" (Boo 2001).

Conclusion

Pokemon and Neopets represent the important changes in children's entertainment that have occurred due to increased marketing via the Internet, the decline of traditional toy licenses, and the dispersion of the children's audience across television channels, from television screens to computer screens, and across games, characters, and stories that span the children and teen markets. Mattel will probably never regain the hold they enjoyed over the children's market from the 1960s into the 1990s, and Toys "R" Us has held on only precariously through the rise of e-commerce and the shifting focus of children from the Saturday-morning television ghetto. Network television competes desperately for advertising revenue against Disney and Nickelodeon, but also the Cartoon Network, the WB, and Fox, so the environment is more fragmented. If these corporate entities have historically viewed children as fickle, they now see Japanese licenses and the pleasures of Web surfing and chatting as further unsettling and unpredictable elements in their work of predicting what children will enjoy. No one in children's entertainment believes that a return is possible to the good old days of mass-market commercials, a network children's block, and television alone as the ruling force in marketing.

In my after-school computer classes at the working-class school Washington and the affluent school Clearview, children formed intense personal connections to Pokemon, Digimon, and Neopets. Virtual pets motivate children to master the skills, including typing, necessary for using the Internet. For children without computers at home, Neopets can even compel them to go to the local public library. As one fourth-grader wrote in an article encouraging others to visit Neopets.com, "Neopets are a lot of fun because you take care of them as if they were real. I like Neopets because whenever I go to the library that's the only thing I want to do." Children rarely know where their toys and media come from; in fact, most of the children in my class never knew Pokemon was Japanese until I told them. Word of mouth about what is fun, however, travels fast, and the Web has intensely accelerated this process, leading the promotions industry to dub the process "word of mouse."

When I first looked at the Neopets site, I was stunned at the embedded advertising, and the commercial audacity of some of the schemes for gaining points. When the children in my class look at the site, they primarily see opportunities for victory, fame, and fortune in a fan community. After some reflection on why our reactions were so different, I realized I should ask the children straight out why they thought Neopets.com existed. To my surprise, all the students gave the same answer, more or less: Neopets was just the cool idea of a lone individual who wanted to share the fun. Neopets existed because somebody, somewhere, had made up something cool, had a good idea, and put it on the Web for our enjoyment. Their image of the creator was a private hobbyist. When I asked if it cost money to produce Neopets, they answered no, just the cost of the computer. Their high level of involvement helped to dull their awareness of the commercialism. As one ratings analyst praised the site, "As a user, you're creating part of the content, and so you feel a personal, emotional connection to it" (Winding 2002). Because Neopets does not involve real money in its online transactions, and there is no membership fee, children find it more difficult to see the process of commercialization.

In the course of my research with children, I have posed the following question to over 150 students at the two elementary schools where I have taught: Is there advertising on Neopets.com? At the working class school, the students' first answer was a simple no: they believed that they encountered no advertising on Neopets. After some prompting, they identified only the flashing banner ads on the log-in page as an example of advertising.

When I challenged students to take a more careful look through the site, trying to find examples of advertising, it became apparent that it was necessary to specify that product mentions were a kind of advertising. Once students grasped this concept, they quickly found dozens of examples of ads on the Neopets site. Often children enjoyed finding advertising as a game, a form of Internet scavenger hunt, and children surprised me with examples of embedded advertising I had never reached on the site. The students are equally enthusiastic about introducing me to dozens of ways to gamble for points.

When children get their hands on computers they generally turn out to be very good at them—because their openness to play with the machines allows them to learn much that eludes adults. Websites such as Neopets.com—like video games themselves only less costly—reward tinkering, strategizing, obsessive play, and absorption. Clearly Neopets has successfully offered children a range of activities and a level of complexity they find compelling, and has probably done much to attract girls to the pleasures of role-playing and simulation games, continuing the process started by Pokemon. But Neopets is not free, and adults have a responsibility to help children understand this fact.

We need to think of new ways to teach children about the Internet and the information economy, because phenomena like Pokemon demonstrate rapid transformations in business approaches to the children's market. Major features of children's media have changed in recent years: target marketing based on gender has eased, Japan has become as reliable a supplier as the United States to Saturday-morning television, and the Internet is fully integrated into the mechanics of children's marketing. Children are pretty savvy about commercial television. On the other hand, ignorance about the commercial nature of the Web among children is rampant, not just because they don't understand product placement, but because they cannot see how monitoring and profiling—data mining—can be a business in and of itself. The most intense marketing activities on the Web occur at a level that is impossible for children to perceive. Neopets' Immersive Advertising makes those activities associated with television (host-selling, deceptive advertising, product placement, and sweepstakes and premiums) seem straightforward. At the level of "free" sites such as Neopets that are easily accessible through public-access sites, children need to understand the full ramifications of brand awareness, retail markups for branded items, and the mercenary uses of fan enthusiasm. Children with the luxury of a high-end residential computer setup need to be aware of how much information about themselves they are leaving behind as they surf the Web, whether through music downloading systems that monitor all Web traffic, through cookies planted on consumers' hard drives, or through customized Web pages. I have learned to make a point of regularly asking children in my class what is abbreviated in the term *dot-com*. So far not a single child has known the answer. Several children responded that *com* stands for community—a completely understandable mistake, given the hype about the nature of human relations on the Web, but a dismaying one.

In my book *Sold Separately*, I criticize media education efforts of the 1980s and 1990s, in their underestimation of children's ability to comprehend television programs and the selling intent of commercials. Many of these programs posited a naïve child viewer and assumed that consumer desire stemmed only from advertiser manipulation (Seiter 1993). In the new mix of children's entertainment, toys, and the World Wide Web, I would strenuously advocate providing children with as much information as possible about product placement on websites and in video games, about sponsored content on the Web, and the use of surveys and sweepstakes to profile consumers and households. This should not be restricted to blatantly commercial sites, but should include those sites where "educational" games or information are sponsored by corporations and special-interest groups. There is much pleasure to be had by children in playing online games, joining fan communities and chat rooms, and trolling the Web for new amusements. Commercialization is more pervasive and more integral than it is in television programming, however, and it would be a grave mistake to leave children with the impression that all they need understand is the equivalence between banner ads and television commercials. Given the shifting, transitory, and multiple nature of the Web and its inventive attempts to make money, this calls for a high level of research and vigilance. The World Wide Web is a more aggressive and stealthy marketer to children than television ever was, and children need as much information about its business practices as teachers and parents can give them.

conclusion

"Off the Treadmill!"

Notes on Teaching Children the Internet

When computers first entered the educational sphere, I can recall the teachers talking to us in elementary school about how we would be able to converse with students in Africa online. Fifteen years later and I have yet to talk to anyone from Africa, in fact I have yet to even talk to someone in Mexico on the web, nor have I seen sites related to the country or its language.

—WASHINGTON TUTOR AND UNIVERSITY STUDENT

. . . to enlarge the possibility of intelligible discourse between people quite different from one another in interest, outlook, wealth, and power, and yet contained in a world where, tumbled as they are into endless connection, it is increasingly difficult to get out of each other's way.

—CLIFFORD GEERTZ, ON THE PURPOSE OF ETHNOGRAPHY (1988, 146)

While inequalities in the distribution of money, education, and resources have only worsened in the last three decades, Americans cling nevertheless to the myth that we are mostly classless. The digital divide was no more than a new name for class difference, but it spurred philanthropy, especially for educational technologies to benefit children in the United States. Closing the gap between technology rich and technology poor seemed easier task to address than the wider and deeper deprivations faced by so many children in our society: supplying hardware and software seemed a simple, attainable, and therefore attractive fix for educational inequalities. While I recognized this contradiction in the digital divide discourse, I felt that any efforts to redistribute resources were at least a good place to start. The move to wire the information poor did a great deal to encourage those with good education and plentiful resources to reach out to neighborhoods they normally would avoid, due to de facto segregation by ethnicity and income in U.S. cities. My project was only one of many that offered academics an opportunity to be useful, to share some of the wealth concentrated at universities. Now the federal government and many other research agencies have lost interest in the digital divide. The digital divide has passed out of fashion at just the moment when a lack of technology access has been recognized and theorized as an intractable social problem, a phenomenon with deeply undemocratic social consequences.

My computer class dramatically demonstrated how speedily Internet skills can be acquired by children, even by those who do not have the benefit of classroom or home use. Yet the longer I have taught this project, the more troubled I have become by the fact that adequate funding in the United States for basic educational needs, such as teachers, books, and classrooms, is hard to come by when so much has been spent (and also donated during the heyday of the Internet economy) to buy children computers. Doubts about the precise motivations of the social agenda to bring the Internet to all began to grow: Was the philanthropic goal that of providing a public service and enhancing public education or was the goal to use the Web to create and reach new markets?

The question is now moot, since funding for projects like mine have dried up. Even affluent school districts struggle to keep up with the enormous costs of upgrading computers and software, as the industry ceaselessly retires software and hardware from the marketplace after a couple of years. Planned obsolescence is the guiding principle of the new technology industries, and educational institutions are poorly situated to bear the costs of constant replacement and upgrading. Families who have not yet purchased a new computer with an Internet connection do not appear likely to take on the added expense now, when over one third of families in San Diego must spend *more than half* their income just to pay the rent (Weisburg 2002).

If the Internet is an educational boondoggle, it also may be impossible to forego now that it is so firmly established as a medium of social interaction. There are clear benefits that technology access can bring to disheartened and disenfranchised student populations. Computer access compensated for some of the obstacles to boys' success in elementary school—especially working-class boys, for whom professional careers are a distant and unrealistic dream. Against overwhelming odds, the boys have been devoted attendees at this noncompulsory class, and this I believe that is an accomplishment in and of itself. These boys find themselves surrounded by female teachers, most of them white. They are more physically active and louder than teachers would like them to be (and more so than the teachers perceive the girl students to be). For such boys, computers can be a hook to do some sustained academic work. Boys' fascination with the computer as a toy motivates them to work harder. Boys who I never would have expected to get off their skateboards long enough to come to an extracurricular class have returned for years to the Washington class drawn by a classroom community in which their popular cul- ture interests and local knowledge were valued. Would they have enrolled in a newspaper writing or photojournalism class? I don't think so. But a computer class held sufficient cachet to bring them in. For working-class boys, the greater auton- omy offered by computer classes is especially important. They can be involved in school without feeling as though they are capitulating to the demands of others. My original research interests focused on assisting girls in gaining an equal com- petence with computer technology, and most of my previous work has been in the realm of feminist cultural studies. My observations at Washington led me to shift my concern to working-class African American and Latino boys, who seem to be the most alienated from the educational experience, and the most at risk of failure.

Using the Internet as a source of information on sports, music, TV, and film drew the Washington students into participating in what was essentially a writing program. In the long run, our lessons about writing and about image making were more valuable than the lessons about computers, but the computers are what got the children there in the first place. The high teacher-student ratio, and the effort to provide attentive and encouraging relationships with the tutors in the after- school class are what kept them there after the novelty wore off.

For parents and teachers in a position to facilitate learning via the Internet, the following recommendations emphasize the importance of reading and of typ- ing as the crucial forms of computer literacy, and the need to examine websites, databases and search engines critically:

1. Demonstrate the cultural and geographic biases of the Internet. Think about the world compared to the world of the Web. Take students to websites originating in other countries and written in other languages. Ask students to reflect on who is missing and who is over-represented, in their own communities, their nations, and their world. What education researcher Maria de la Luz Reyes advises is doubly important for teaching on the Web: teachers must constantly look for ways to demonstrate "the value of cultural and linguistic diversity in concrete and authentic ways" (Reyes 1992, 432).

2. Recognize the class bias in assumptions about computer and Internet access. Teachers should not require or especially reward word processed papers of elementary-school children. Visiting the public library to do research in books is at least as valuable as surfing the Internet, even if it does not yield the same number of color illustrations. Many schools and teachers, along with countless government agencies, have begun to use e- mail and the Internet as forms of communication with parents and chil-

dren. This needs to be reevaluated, and stopped. Even if one child in the class is "un-wired," this places a prohibitive burden on the family who must repeatedly identify themselves as non–computer users.

3. Familiarize students with reliable, nonprofit sources of information on the Internet, such as the search engines of university libraries, nonprofit organizations, and alternative news organizations. Explain the workings of Internet filtering devices, as installed by many school districts and all libraries receiving federal funding, and the arbitrariness of these systems.

4. Teach children about the economics of the Internet, and the business of search engines. Explain about the dual selling properties of the Web: trying to get you to purchase a product, on the one hand, and trying to get you to divulge information about yourself as a potential consumer, that can be sold to marketers, on the other. As data mining and profiling have taken over, these more hidden forms of commercialism have to be explained slowly and carefully to children, not just as a privacy issue, but as a basic education in business economics. Even struggling readers in my class often found it easier to make it through a *Wall Street Journal* article than something from the local newspaper. Assigned reading from the business press can rapidly open children's eyes to the profit motives behind the Internet (for example, searchenginewatch.com).

5. Introduce children to the genres and conventions of different forms of writing that are commonly found on the Internet. Children need explicit instruction in the distinctive nature and purposes of public relations material, and the difference between publicity and "purer" forms of information.

6. Explain the Web search as a process that requires continual refinement, reflection, and innovation. Careful checks of spelling are necessary: this is often made easier by pairing children up in front of the screen to act as spell checkers. Try different search engines frequently and compare the information they yield. Have the children play detective to try to uncover what is sponsored content and who "authored" it.

7. Make the physical act of reading Web pages easier for children by enlarging fonts, and paring down dense websites. Require reading, not just browsing, but recognize how discouraging a screen full of small text can be for struggling readers. Check to make sure children understand—and remember—computer terminology. Teach and reteach concepts such as cursor, browser, search engine, and URL.

8. Assign Internet scavenger hunts in which students can compare answers to the same question, and analyze how sponsored content, product placement, and the sale of goods and services via the Internet affect the outcome of searches.

9. Get off the Internet and spend time using software that is better suited to children's ages and levels, such as Appleworks, Kid Pix, and Mavis Beacon Teaches Typing. Often children have only scratched the surface of these programs in their introduction to them, even if they have specialized computer instruction. Learning one program in depth is much more instructive in the long run about the possibilities of computing than limitless hours surfing the Web. Better yet, get offline and provide face-to-face contact with real people, in workplaces and libraries, and at events. Give students opportunities to question and engage in dialogue with knowledgeable adults, and value oral communication as well as computer-mediated.

10. Skip the software upgrades, if the computers children are using are working fine. New software releases are often full of bugs. It is a time-consuming and sometimes futile process to restore things to working order after upgrading. Remember that each software change can require hundreds of dollars in hidden costs for printers, cables, and other software upgrades, on top of dozens of staff hours in identifying bugs, downloading new drivers and patches, and troubleshooting the systems. Children rarely employ a fraction of the computing power of the machines they use. Bear in mind the slogan of one group of programmers and computer geeks who restore old computers and game systems: "Off the treadmill!"

In closing, I am well aware that much ethnographic writing "consists of incorrigible assertion," as Clifford Geertz puts it (Geertz 1998, 6). Yet when university professors teach in elementary school classrooms they engage more deeply with the everyday restrictions placed on public school teachers. While I have been critical of some individual teachers in my observations here, most of the teachers I saw at Washington were dedicated and deeply empathetic to the students. They often spent thousands of dollars of their own money to improve the learning experience of the students. When enrichment activities such as my after-school class become available, these teachers went out of their way to contact parents to inform them of available extracurricular activities. The legacy of action research is to expand the children's sense of the possibilities, the rewards and the experiences that await them if they can stay the course and make it to college. Professors can learn a tremendous amount from extended and extensive contact with young students and the public schools in which they struggle. Such work can have an invigorating effect on academic theories of ideology, of identity, and of role of media and technology in everyday life.

I have many reservations about the Internet as a highly commercialized information source, and a uniquely aggressive tool for marketing to children. Similarly, the high rate of computer obsolescence, the difficult work of replacement, and the need for constant technical support are all huge expenses for schools. Now that the initial enthusiasm for school computing has been tempered by the sobering realities of shrinking budgets, it is a good time to reevaluate priorities for school spending.

More and better-paid teachers will always be more educationally beneficial than computer technology from an educational standpoint. The process of education cannot be made more cost-effective through technology. In fact, education cannot be measured in terms of efficiency, and quantifiable results; back-to-basics school reform has led to more alienated students and more corrupt school administrators. If the last two decades of research on technology in the schools have taught us anything, it is that computers, without constant adult supervision, are fundamentally unsuited to children's needs. Teachers can meet Geertz's challenge to "enlarge the possibility of intelligible discourse between people quite different from one another" (Geertz 1998, 5). Teachers must educate children to be critical thinkers about the Internet.

appendix

Digital Media Pedagogy

Digital Media Pedagogy is a seminar devoted to studying the best techniques for teaching digital media: computer software such as Word, Photoshop, and Premiere, digital still cameras, and layout. We will explore our topic through alternating lecture discussion periods and hands-on teaching in the computer lab. What are the special challenges digital media present to teachers and students? How do digital media compare to older technologies such as typewriters, film cameras, and analog video? How do gender, class, and age affect the way students and teachers respond to and learn digital media? Are computers necessary for good elementary school education? Students are required to do at least 10 hours of fieldwork and teaching assistance at Washington Elementary School.

Discussion questions and writing prompts:

Find two advertisements for computers and describe, analyze, and compare their representations of gender. How are the technologies characterized? How is the experience of using the computer characterized? Who is the target audience?

Describe your experiences learning word processing software. What do you remember most about it? Under what conditions did you learn it? How would you evaluate your learning outcome and what would you generalize from this experience about the most crucial factors in teaching? Please be as specific and detailed as possible in describing your experiences. Try to bring into your discussion the role that gender, access, and the home-school connection might have played.

Compare your experiences learning word processing to those learning Photoshop. Which kind of program was easier to learn and why? How did Photoshop prepare you for other graphics programs? What problems have you noticed the students at Washington have with word processing compared to Kid Pix and how would you account for these?

Find two newspaper articles that discuss new technologies in a school or classroom. How are the technologies characterized? Who paid for them?

Are any of the drawbacks Clifford Stoll (1999) describes brought into the news story? How are the students, teachers, and/or computer experts characterized? How do the news stories explain the virtues and probable uses of computers in a classroom? Are you skeptical of anything in the article? Why or why not?

Describe your interactions around the computers with the students at Washington Elementary. What do the kids seem most and least interested in? How would you explain their interests and abilities? Did you notice any of the same gender dynamics discussed by Huber and Schofield (1998)? How do the Washington children seem similar to and different from the European children described in Sonia Livingstone's book *Young People and New Media* (2002)?

Compare your experiences in the university computer teaching lab and at the Washington computer lab. What did you see that was comparable in terms of gender differences? What differed? Please try to describe your observations in as much detail as possible. Do you think you have been strongly aware of gender stereotyping vis-à-vis computer use in your socialization with these machines?

What educational technologies were used when you were in fourth grade (what you wrote with and on, what you used to look at things, what you read from)? Which ones did you feel closest to and why? Which ones did you struggle with? Now ask someone over forty these same questions. Compare and contrast these experiences with what you expect to find in a fourth-grade classroom today.

Many adults fear that the computer is a solitary activity for children, especially when used in their bedrooms. On the other hand, some scholars have noted that computer games encourage group involvement through the sharing of tips and cheats, and information circulating through networks of friends, more often than it is read in a specialty magazine. Do you see children's computer use as primarily social or as solitary?

Describe the relations to and uses of computers of your parents and siblings. Then attempt to explain each family member in terms of computer liking and computer anxiety. Place your family's experiences in terms of social and kin networks (extended family or nuclear family); education, work experience; household responsibilities; native language, gender, class; and any other social-contextual factors you think helps to explain the variation in usage patterns.

Develop a lesson for the Washington students that introduces them to the Internet by taking them through visits to three Spanish-language websites. Try to find websites that would be especially appealing or interesting to the students, and develop some activity for them to do at each site.

Describe your interactions around the computers with the students at Washington Elementary. What do the kids seem most and least interested in? What has surprised you about their abilities compared to those of your peers? How would you describe their interests and abilities? What would you use to explain the differences among the students, besides gender?

Bibliography

Allison, Anne. 2004. "Cuteness as Japan's millennial product." In *Pikachu's Global Adventure: The Rise and Fall of Pokemon*, ed. Joseph Tobin. Durham, NC: Duke University Press.

———. 2000. *Permitted and Prohibited Desires: Mothers, Comics and Censorship in Japan*. Berkeley: University of California Press. (Original publication: Boulder, CO: Westview Press, 1996.)

Apple, Michael. 1996. *Cultural Politics and Education*. New York: Teachers College Press.

———. 1995. *Education and Power* (2d edition). New York: Routledge.

———. 1993. *Official Knowledge: Democratic Education in a Conservative Age*. New York: Routledge.

Barthes, Roland. 1982 [1972]. "The World of Wrestling." In *A Barthes Reader*, ed. Susan Sontag. New York: Hill and Wang.

Bauder, Don. 2001. "Depressed tourism adds to ballpark woes." *San Diego Union Tribune*, 23 September: D1. Available at: www.traveltax.msu.edu/news/Stories/sandiego_ut13.htm.

Berliner, D., and B. Biddle. 1995. *The Manufactured Crisis: Myths, Fraud, and the Attack on America's Public Schools*. Reading, MA: Addison-Wesley.

Bloch, L. and Lemish, D. 1999. "Disposable love: The rise and fall of a virtual pet." *New Media and Society* 1, no. 3: 283–303.

Boo, Kristi. 2001. "Online pet world keen on Singapore base." *Asiaweek*, 12 June.

Bourdieu, Pierre. 2000. *Pascalian Meditations*. Palo Alto, CA: Stanford University Press.

Brandt, Stacy. 2002. "Who was that masked man?" Exploring the incredibly strange world of excess that is lucha libre." *Daily Aztec* (San Diego State University newspaper). 5 December. p. 9.

Bridis, Ted. 2001. "Bush staff wants to slash programs set up to help close 'digital divide.'" *Wall Street Journal*, 15 February. http//online.wsj.com/article/0,SB982190280545779663.djm,00.html

Buckingham, David. 2002. *Small Screens: Television for Children*. London: Leicester University Press.

———. 2000. *The Making of Citizens: Young People, News and Politics*. London: Routledge.

Cassells, Justine, and Henry Jenkins. 1998. *From Barbie to Mortal Kombat: Gender and Computer Games*. Cambridge, MA: MIT Press.

Chakravartty, Paula. 2004. "Telecom, national development and the Indian state: A postcolonial critique." *Media, Culture and Society* 26, no. 2: 227–40.

Clark, Roy Peter. 1987. *Free to Write: A Journalist Teaches Young Writers*. Portsmouth, NH: Heinemann.

Cobo, Leila. 2002. "WB hopes everyone gets 'Lucha.'" *Billboard* 114, no. 32 (August 10): 33.

Cockburn, Cynthia. 1999. "The material of male power." In *The Social Shaping of Technology*, ed. Donald MacKenize and Judy Wajcman. Buckingham: Open University Press.

Cole, Michael. 1996. *Cultural Psychology: A Once and Future Discipline*. Cambridge, MA: Belknap Press of Harvard University Press.

Cole, Michael, Yrgo Engestrom, and Olga Vasquez, eds. 1997. *Mind, Culture and Activity: Seminal Papers from the Laboratory of Comparative Human Cognitio*. Cambridge: Cambridge University Press.

Cole, Michael, and James V. Wertsch. 1996. *Contemporary Implications of Vygotsky and Luria*. Worcester, MA: Clark University Press.

Cuban, Larry. 2001. *Oversold and Underused: Computers in the Classroom*. Cambridge, MA: Harvard University Press.

Cummins, Jim, and Dennis Sayers. 1997. *Brave New Schools: Challenging Cultural Literacy*. New York: St. Martin's Griffin.

Dennis, Geoff. 2002. "Internet: Missed opps beyond the banner." *Strategy*, 6 May, p. 14.

Dodge, John. 1999. "The Pokemom's [sic] shopping secret: Turning to the World Wide Web." *Wall Street Journal: Interactive Edition*, 10 August.

Dyson, Anne Haas. 1997. *Writing Superheroes: Contemporary Childhood, Popular Culture, and Classroom Literacy*. New York: Teachers College Press.

The Economist. 1999. "Latinos in Silicon Valley: The digital Divide." 17 April, p. 53.

Electronic Gaming Monthly. 1999. "What's the deal with Pokemon?" October, pp. 163–72.

Fagot, Beverly I., Mary D. Leinbach, and Richard Hagan. 1986. "Gender labeling and adoption of sex-typed behaviors." *Developmental Psycholosgy* 22, no. 4: p. 440–443.

Fisherkeller, JoEllen. 2002. *Growing up with Television: Everyday Learning Among Young Adolescents*. Philadelphia, PA: Temple University Press.

Foulkes, Nicholas. 2001. "Pocket monsters: Tamagotchi die, but can Pokemon go for ever?" *Financial Times*, 5 May, p. FT 24.

Freire, Paolo. 1994. *Pedagogy of Hope: Reliving Pedagogy of the Oppressed*. New York: Continuum.

Geertz, C. 1988. *Works and Lives: The Anthropologist as Author*. Stanford, CA: Stanford University Press.

Haddon, Leslie. 1992. "Explaining ICT consumption: The case of the home computer." In *Consuming Technologies: Media and Information in Domestic Spaces*, ed. Roger Silverstone and Eric Hirsch. London: Routledge.

———. 1988. "Electronic and computer games: The history of an interactive medium. *Media, Culture and Society* 29, no. 2: 52–75.

Harris, Kathryn. 1993. "Taking a megabyte of the market." *Los Angeles Times*, 18 March, p. D1.

Harrison, A. W., R. K. Rainer, Jr., and W. A. Hochwarter. 1997. "Gender differences in computing activities." *Journal of Social Behavior and Personality* 12, no. 4: 849–68.

Harrison, A. W., R. K. Rainer Jr., W.A. Hochwarter, and K. R. Thompson. 1997. "Testing the self-efficacy-performance linkage of social-cognitive theory." *Journal of Social Psychology* 137, no. 1: 79–87.

Hattori, Fumiko. 2001. "Japan press: Mobile games fare better than fixed game gear." *Wall Street Journal: Interactive Edition*, 8 March.

Hill, Jane H. 1998. "Language, race and white public space." *American Anthropologist* 100, no. 3: 680–89.

Hoffman, Jeanette. 1999. "Writers, texts and writing acts: Gendered user images in word processing software." In *The Social Shaping of Technology* (2d edition), ed. Donald MacKenzie and Judy Wajcman. Buckingham: Open University Press.

Huber, B. R., and J. W. Schofield. 1998. "Gender and the sociocultural context of computing in Costa Rica." In *Education/Technology/Power: Educational Computing as a Social Practice*, ed. Hank Bromley and Michael W. Apple. Albany: State University of New York Press.

Internet World. 2001. 1 May.

Iwabuchi, Koichi. 2004. "How 'Japanese' is Pokemon." In *Pikachu's Global Adventure: The Rise and Fall of Pokemon*, ed. Joseph Tobin. Durham, NC: Duke University Press.

Jackson, David (1998) "Breaking out of the binary trap: Boys' underachievement, schooling and gender realtions." In *Failing Boys? Issues in Gender and Achievement*, ed. Debbie Epstein, Jannette Elwood, Valerie Hey, and Janet Maw. Buckingham: Open University Press.

Jenkins, Henry. 1997. "'Never trust a snake': WWF wrestling as a masculine melodrama." In *Out of Bounds: Sports, Media, and the Politics of Identity*, ed. Aaron Baker and Todd Boyd. Bloomington: Indiana University Press.

Jervis, Kathy. 1996. "'How come there are no brothers on that list?' Hearing the hard questions all children ask." *Harvard Education Review* 66, no. 3: 556–49.

Jessen, Carsten. 2000. "Girls, boys and the computer." Available at: www.hum.odu.dk/center/kultur/GBC.htm. Accessed 14 November.

———. 1996. "Girls, boys and computers in the kindergarten: When the computer is turned into a toy." In *Children's Computer Culture: Three Essays on Children and Computers*, Working Paper 8, Child and Youth Culture, Department of Contemporary Cultural Studies, Odense University. Odense: University of Southern Denmark.

Kafai, Yasmin. 1998. "Video game design by girls and boys: Variability and consistency of gender differences." In *From Barbie to Mortal Kombat: Gender and Computer Games*, ed. Justine Cassells and Henry Jenkins. Cambridge, MA: MIT Press.

Katsuno, Mifune, and Jeffrey Maret. 2004. "Localizing the Pokemon TV series for the American market." In *Pikachu's Global Adventure: The Rise and Fall of Pokemon*, ed. Joseph Tobin. Durham, NC: Duke University Press.

Keefe, Bob. 2002. "Blacks confront a digital divide." *Atlanta Journal-Constitution*. 10 March. Business, p. 1P.

Kinder, Marsha. 1991. *Playing with Power in Movies, Television and Videogames from Muppet Babies to Teenage Mutant Ninja Turtles*. Berkeley: University of California Press.

Kohn, Alfie. 2000. *The Case Against Standardized Testing: Raising the Scores, Ruining the Schools*. Portsmouth, NH: Heinemann.

Kramarae, Cheris. 1988. *Technology and Women's Voices: Keeping in Touch*. New York and London: Routledge and Kegan Paul.

Krendl, K. A., M. C. Broihier, and C. Fleetwood. 1989. "Children and Computers: Do Sex-Related Differences Persist?" *Journal of Communication* 39, no. 3: 85–93.

Layden, Joseph. 2000. *The Rock Says: The Most Electrifying Man in sports-Entertainment*. New York: Reagan Books.

Levi, Heather. 2001. "Masked media: The adventures of Lucha Libre on the small screen." In *Fragments of a Golden Age: The Politics of Culture in Mexico*, ed. Gilbert M. Joseph, Anne Rubenstein, Eric Zolov, and Elena Poniatowska. Durham, NC: Duke University Press.

———. 1999. "On Mexican pro wrestling: Sport as melodrama." In *SportCult*, ed. Randy Martin and Toby Miller. Minneapolis: University of Minnesota Press.

Livingstone, Sonia. 2002. *Young People and New Media: Childhood and the Changing Media Environment*. London: Sage.

Lowney, Kathleen S. 2003. "Wrestling with criticism: The World Wrestling Federation's ironic campaign against the Parents Television Council." *Symbolic Interaction* 26, no. 3: 427–446.

Margolis, Jane, and Allan Fisher. 2002. *Unlocking the Clubhouse: Women in Computing*. Cambridge, MA: MIT Press.

McChesney, Robert. 1997. *Corporate Media and the Threat to Democracy*. New York: Seven Stories.

———. 1996. "The Internet and U.S. communication policy-making in historical and critical perspective." *Journal of Communication* 46, no. 1: 98–124.

McChesney, Robert, and Ben Scott. 2004. *Our Unfree Press: 100 Years of Media Criticism*. New York: New Press.

McNeil, L. 2000. *Contradictions of School Reform: Educational Costs of Standardized Testing*. London: Routledge.

Miller, Toby. 2000. *Sportsex*. Philadelphia, PA: Temple University Press.

Milliot, Jim. 2004. "Jovanovich returns to Pearson." *Publishers Weekly* 251, no. 22 (May 31): 10.

Milone, Michael. 1999. "Enterprise computing goes to school." *Technology and Learning* 20, no. 2: 28.

Mitchell, Claudia, and Jacqueline Reid-Walsh. 2002. *Researching Children's Popular Culture: The Cultural Spaces of Childhood*. New York: Routledge.

Mitra, Sugata. 2003. "Minimally invasive education: A progress report on the 'hole-in-the-wall' experiments." *British Journal of Educational Technology* 34, no. 3: 367–71.

———. 2000. "Children and the Internet: New paradigms for development in the 21st century." Keynote address, Asian Science and Technology Conference, Tokyo, Japan. 6 June.

Mitra, Sugata, and Vivek Rana. 2001. Children and the Internet: Experiments with minimally invasive education in India. *British Journal of Educational Technology* 32, no. 2: 221–32.

Mosco, V. 1998. "Myth-ing links: power and community on the information highway." *Information Society* 14, no. 1: 57–62.

NCS Learn Home Page. 2002. Available at: www.ncslearn.dom/press/company/html. Accessed 2 February.

NeoPets Privacy Policy. 2001. Available at: http://www.neopets.com/privacy/phtml. Accessed 6 November.

Nevin, Alan. 2001. "East Village vitalization." *San Diego Metropolitan* May: Real Property. Available at: www.sandiegometro.com/2001/may/property.html.

NIIT Technologies Ltd., 2005 "NIIT Ltd, 1ar Quarter '05 Consolidated Results" http://www.niit.com/niit/ContentAdmin/NWS/NWSPR/NWSPR5/pr-270704-ltd-results.htm

Noble, David. 2001. *Digital Diploma Mills: The Automation of Higher Education*. New York: Monthly Review Press.

———. 1984. *Forces of Production: A Social History of Industrial Automation*. New York: Knopf.

Pearson Education. 2002. Available at: www.pearsoned.com; www.perasoned.co.uk.

PR Newswire. 2002. "Although television still reigns supreme in advertising effectiveness Internet tapped as favorite media by American youth, finds Neo Pets." 18 March.

Reyes, Maria de la Luz. 1992. "Challenging venerable assumptions: Literacy instruction for linguistically different students." 62, no. 4: 427–446.

Rommelman, Nancy. 2002. "Just Who Are Those Masked Men?" *Los Angeles Times*. Calendar Weekend, Part 6, p. 36, 22 August.

Rotenberg, Marc. 2002. "Testimony and statement for the record on Consumer Privacy Protection Act of 2002." Subcommittee on Commerce, Trade and Consumer Protection, Epic Privacy Information Center, 24 September. Available at: www.epic.org/privacy/consumer/hr4678testimony_92402.html.

Saltman, Kenneth. 2002. "Junk-king education." *Cultural Studies* 16, no. 2: 253.

Saunders, Terry McNeill. 1998. "Play, performance and professional wrestling: An examination of a modern day spectacle of absurdity." Ph.D. thesis, University of California, Los Angeles.

Schiller, Dan. 2000. *Digital Capitalism: Networking the Global Marketing System*. Cambridge, MA: MIT Press.

Schofield, Janet Ward. 1995. *Computers and Classroom Culture*. Cambridge: Cambridge University Press.

Schofield, Janet Ward, and Ann Locke Davidson. 2002. *Bringing the Internet to School: Lessons from an Urban District*. San Francisco, CA: Jossey-Bass.

Scott, Ben. "A contemporary history of digital journalism." *Television and New Media*. Forthcoming.

Seiter, Ellen. 1999. *Television and New Media Audiences*. Cambridge: Oxford University Press.

———. 1993. *Sold Separately: Children and Parents in Consumer Culture*. New Brunswick, NJ: Rutgers University Press.

Silverstone, Roger, and Eric Hirsch. 1992. *Consuming Technologies: Media and Information in Domestic Spaces*. London: Routledge.

Smackdown. 2003. Available at: smackdown.wwe.com/superstars/mysterio/index.html. Accessed December 10.

Stanley, Laura. 2001. *Beyond Access: Qualifying the Digital Divide*. UCSD Civic Collaborative, Unpublished paper.

Stoll, Clifford. 1999. *High-Tech Heretic: Why Computers Don't Belong in the Classroom and Other Reflections by a Computer Contrarian*. New York: Doubleday.

Tam, Pauline. 2001. "Neopet(ty) crimes on the Internet: How hackers invaded an online kids game, and scammed unsuspecting players." *Ottawa Citizen*, Tech Weekly, 30 July, p. B3.

Thomas, Angela, and Valerie Walkerdine. 2002. "Girls and computer games." Available at www.woman.it/cyberarchive/files/thomas.htm. Accessed 16 February.

Thompson, Audrey. 1998. "Not the color purple: Black feminist lessons for educational caring." *Harvard Educational Review* 68, no. 4: 522.

Tomsho, Robert. 2002. "Children's access to technology still affected by income and race." *Wall Street Journal*, 5 July 2002. http://online.wsj.com/article/,,SB1025879996213121280.djm.)).htm.

Turkle, S. 1995. *Life on the Screen: Identity in the Age of the Internet*. New York: Simon and Schuster.

———. 1984. *The Second Self*. New York: Simon and Schuster.

Valenzuela, Angela. 1999. *Subtractive Schooling: US-Mexican Youth and the Politics of Caring*. Albany: State University of New York Press.

van Zoonen, Liesbet. 1992. "Feminist theory and information technology." *Media, Culture and Society* 14, no. 1: 9–29.

Vasquez, Olga. 2002. *La Clase Magica: Imagining Optimal Possibilities in a Bilingual Community of Learners*. Mahwah, NJ: Lawrence Erlbaum.

Vasquez, Olga, Lucinda Pease-Alvarez, and Sheila M. Shannon. 1994. *Pushing Boundaries: Language and Culture in a Mexicano Community*. Cambridge: Cambridge University Press.

Walkerdine, Valerie. 1999. "violent boys and precocious girls: Regulating childhood at the end of the millennium." *Contemporary Issues in Early Childhood* 1, no. 1: 3–23.

Wall Street Journal. 2001. "Low-income students are less likely to have internet access, report finds." May 10. http://online.wsj.com/article/0,,SB989458506916171291.djm,00.html

Wall Street Journal. 2001. "Nintendo Co. Ltd." Cita's Briefing Books.

Warschauer, Mark. 2003. *Technology and Social Inclusion: Rethinking the Digital Divide*. Cambridge, MA: MIT Press.

Weber, Jonathan. 1993. "Turning the pages: Publishing an overlooked gold mine for Paramount." *Los Angeles Times*, 2 October, p. D1.

Weingarten, Marc. 2002 "As children adopt pets, a game adopts them." *New York Times* 21 February, D5.

Weisberg, Lori. 2002. "Local housing costs drain family budget." *San Diego Union Tribune*, 27 August, p. 1.

Willett, Rebekah. 2004. "The multiple identities of Pokemon fans." In *Pikachu's Global Adventure: The Rise and Fall of Pokemon*, ed. Joseph Tobin. Durham, NC: Duke University Press.

Willett, Rebekah. 2003. "Living and learning in chatrooms (or does informal learning have anything to teach us?)" *Education et Societes*, vol. 2. p. 12.

Winding, Elizabeth. 2002. "Immersed in child's play." *Financial Times*, 10 June, p. 17.

Yano, Christine. 2002. "Kitty litter: Consuming Japanese cute." Paper presented at the 3rd World Congress on Toy Research, Toys, Media and Games, University of London, 19 August.

Index

Toby Miller
General Editor

Popular Culture and Everyday Life is the new place for critical books in cultural studies. The series stresses multiple theoretical, political, and methodological approaches to commodity culture and lived experience by borrowing from sociological, anthropological, and textual disciplines. Each volume develops a critical understanding of a key topic in the area through a combination of thorough literature review, original research, and a student-reader orientation. The series consists of three types of books: single-authored monographs, readers of existing classic essays, and new companion volumes of papers on central topics. Fields to be covered include: fashion, sport, shopping, therapy, religion, food and drink, youth, music, cultural policy, popular literature, performance, education, queer theory, race, gender, and class.

For additional information about this series or for the submission of manuscripts, please contact:

Toby Miller
Department of Cinema Studies
New York University
721 Broadway, Room 600
New York, New York 10003

To order other books in this series, please contact our Customer Service Department:

(800) 770-LANG (within the U.S.)
(212) 647-7706 (outside the U.S.)
(212) 647-7707 FAX

Or browse online by series: